Technology Assessment for State and Local Government

Technology Assessment for State and Local Government

A Guide to Decision Making

Lawrence P. O'Keefe

amacom American Management Associations

Library of Congress Cataloging in Publication Data

O'Keefe, Lawrence P.
    Technology assessment for state and local government.

    Includes index.
    1. Technology assessment. I. Title.
T174.5.038   1982   303.4'83   82-71314
ISBN 0-8144-5595-6   AACR2

First Printing

To Helen, Larry, Heather, and Daniel

Contents

1 Introduction 1
2 Technology and Society 11
3 Technology Assessment at the Local Level 25
4 Forecasting and Assessment Tools 41
5 Normative Methods 60
6 Analogies 80
7 Models 97
8 Driving Forces in the 1980s—The Environment 116
9 Driving Forces in the 1980s—Energy 140
10 The Revolution in Electronics 160
11 Putting It All Together 185
  Index 203

Technology Assessment for State and Local Government

CHAPTER 1

Introduction

This book grew out of my firsthand exposure to local governments. Trained as an engineer, I spent several years in a high-technology industry focusing on technical and administrative tasks associated with producing one system or product for a single customer. In this highly technical environment, most of the activity was focused on the problems associated with designing, manufacturing, and delivering a product for a predetermined market.

A stint as a consultant followed, giving me my first experience with local governments. I received some exposure to the environment in which decisions are made but, again, most of my effort was confined to delivering a product that had been contracted for. From there I had an opportunity to join a nationwide technology transfer program sponsored by the National Science Foundation. This program was developed to test the theory that a locally accessible technology transfer person who operated much in the role of the highly successful agricultural extension agent could stimulate the transfer of technology from other sectors of the economy into local governments. The program was established nationally with 27 so-called technology agents placed in 27 cities and counties around the country. The technology agents were made an integral part of the government structure, usually reporting to a mayor or city manager, and had the resources of some major research organizations and high-technology corporations to draw from.

The program was highly successful and still exists today, but with drastically reduced federal funding. It is expensive to fund this local technology presence, and, in retrospect, it is not a concept that could be reasonably extended into every city with federal funds, despite its successes. The reason it was not expanded was because somehow it did not produce what was wanted, not because it was unsuccessful. As I have seen over and over and over again, in a large complex social–political system, technology is but one of a diverse set of issues that must be dealt with if we are to move a product, concept, or process across social and institutional boundaries. During the process of transfer, a whole succession of agenda items—some pure and grand, some unbelievably petty and trivial—must be addressed and technology agents find themselves not so much researching technology as measuring people and attitudes, pacing progress, building up momentum, and patching up defeats.

The manager or administrator who functions in our complex society must constantly come to grips with technology—whatever that may be. If he is having something built, or if he is ordering a large or small off-the-shelf product, or a system or process, or even developing ways of doing things, he has an array of choices. As an administrator, he has an understanding of people, social processes, and decision making. He may even have a specialized technical or nontechnical background from which he evolved. However, particularly as the cost and lifetime of a piece of equipment, a process, or a service increases, an administrator is faced with increasingly sophisticated sellers, operators, and users whose basic agenda may or may not coincide with his own. In short, he finds himself in a very complex decision-making process that involves special skills and tools.

The technically trained person who enters this arena of greater choice also finds a set of new and unfamiliar rules. In-depth technical knowledge must give way to getting something reasonable done in a complex environment that has none of the simple, easy-to-define rules that govern the design and operation of an engineered product or system. In fact, technical considerations are only one part of a complex technical–social–political system. Somewhere there is a need to merge these technical and social considerations to produce an ordered process for making decisions in a multidimensional environment. Over the years, technology transfer organizations and multitudes of outside helpers have tried to assist local government administrators and decision makers in this activity. By and large they have just not been able to do it. By definition these helper organizations are staffed with mobile, quick-witted, and ambitious people. The organizations, originally charted to help their constituencies, seem to develop a life of their own. They focus on expanding, capturing a larger share of their marketplace, or opening up new markets at the expense of the organizations they had hoped to serve. Ultimately, they puff up, wither, and die, and it is left to the localities—the cities and counties, towns and villages, parishes and districts—to help themselves. It is for that purpose that this book is written.

The technology agent has a unique perspective. He is at once an actor and a viewer. He is a part of the organization but somehow different. He is a wild card, a deliberate force for change, a planned disturbance in an organization that emphasizes harmony, discipline, and following the rules. If his role is properly structured, he does not spend much time in the day-to-day operations of a local government but has the opportunity to measure firsthand the operations and the decision-making process in a community. The abrupt change from a world that emphasizes products to a world that emphasizes political processes and service delivery provides a unique opportunity for a technically trained person to view a local government through different eyes. This book is written to reflect that perspective as well.

The opportunity to teach a course in technology forecasting in a master's degree program in public administration brought another opportunity to compare theory with experience, and out of that grew the conviction that technology assessment—measuring the impact of existing technology or the consequences of future technology—is a discipline far more suited to the local government environment than is the process of long-range forecasting. The decision was made, and my experiences as a technology agent, formalized for the classroom, formed the basis for this book.

The book is not intended to be a detailed treatise on the methods and techniques of technology forecasting and assessment;

there are enough reference texts to take readers as far as they care to go in those areas. Nor is it intended to be an elaborate discussion of public policy or the sociology of technology transfer; from an engineer's point of view, there is perhaps too much written on those subjects already. Rather, the attempt is to focus on bridging—to expose the technician to the situation and constraints that surround the measurement and selection of technology in the public arena, and to expose the generalist to some tools that can be used to good advantage in obtaining the benefits of technology while avoiding the pitfalls. So the book is intentionally wider than it is deep; it is intended to convey a perspective more than an in-depth knowledge of the subject of technology assessment.

BOOK ORGANIZATION

The book is broken into four sections: background information, assessment and forecasting techniques, driving forces in the 1980s, and putting it all together. The background section, which contains this chapter and Chapters 2 and 3, provides the framework and background for technology assessment. Chapters 4, 5, 6, and 7 describe the types of assessment tools that can be used. Chapters 8, 9, and 10 discuss energy, the environment, and the revolution in electronics as significant technology drivers in the 1980s. Chapter 11 discusses how technology assessment can be integrated into local government operations and gives an example of its use.

This chapter explains the approach to the book and how it came to be. Chapter 2 describes the intimate involvement of technology and society, using the automobile as an example of technology that brought substantial undesirable as well as desirable changes to our society. It explains how technology forecasting grew up as an adjunct to our military planning processes but clearly has a place in the measurement, planning, and evaluation of our social systems. Technology assessment is a companion to technology forecasting. Where forecasting focuses on the path of progress, assessment focuses on the consequences of that progress. Although separable, the two are somewhat bound together. Any system, physical or social, responds to change, and that very response positively or negatively affects the path of future change. No sensitive forecast can be made without assessing the consequences of change, and no assessment can be made without some knowledge of the path of progress. In Chapter 2, forecasting and assessment are shown to be integral parts of any planning process.

Chapter 3 characterizes local governments as large-scale public service delivery corporations that have a diverse role in our society. In delivering these services, local governments are substantial users of technology. This is clearly evident in the traditional physical services they provide. One has only to think about water supply, wastewater treatment, refuse collection and disposal, roads, bridges, harbors, airports, and other facilities in order to see the link. It also shows up in the regulation of the use of land and the construction of structures.

In the fast-growing area of human service delivery, the uses for technology may be subtle, but they are built in nevertheless. Facilities are required to support these services. Transportation and communications are required. And in the fiscally stringent environment of the 1980s, much stress will be placed on the use of technology to maximize the effectiveness and efficiency of human service delivery. Local governments as social–political organizations have some built-in barriers to the introduction of new technology. Chapter 3 discusses the focus on land use, pressures from above and below, the two-year election cycle, organizational structure, and the fish bowl environment as barriers to innovation. Nevertheless, local governments have a substantial and growing place in our economy and society, and as their role grows, it becomes increasingly more critical to ensure that they effectively use the technologies at their disposal.

Chapter 3 also makes the statement that technology forecasting is a necessary function at the federal level and perhaps the state level, but that as we shift to the localities where problems are less global and more immediate, technology assessment should have a dominant role.

The next four chapters focus on assessment and forecasting tools. Chapter 4 discusses predictive tools, which are more adaptable for forecasting purposes. It describes some of the pitfalls of informal forecasting and points up the need to be prepared for change. Classic predictive tools, which extrapolate future progress from past results, are described along with a discussion of key parameters for prediction and the limitations of extrapolation. The Delphi process, a less rigid, broader-based tool for forecasting in a complex situation is discussed, and my own use of Delphi in an academic setting is illustrated.

Chapter 5 discusses descriptive tools, which by their nature are more adaptable to the assessment process. Relevance trees, morphological models, and flow diagrams are applied in a local government setting to demonstrate how they can be used. The relevance tree, which charts hierarchical structure, is applied to a building inspection department to chart the organization and set priorities. A morphological model is used to define the major elements in a refuse collection vehicle and to outline choices for wastewater treatment processes. The flow diagram, which is sensitive to time-related systems, is applied to an emergency communications system and used to delineate the modes of travel to proposed community service centers.

Chapter 6 presents analogies, a more intuitive but powerful and familiar approach to assessment. Analogies permit the comparison of a present or future situation with a known past situation, allowing us to think in terms of specific differences rather than to try to absorb the whole content of a situation. Thinking by analogy is a common, often used process, and this chapter adds the formal dimension necessary to provide this tool with the structure that makes it powerful and comprehensive.

With the advent of the digital computer, and the supporting computational sciences, simulation or modeling of physical systems came of age. Using the laws of science, we can simulate anything from a chemical reaction to the flight of an airplane, expanding or compressing time as needed and sidestepping the need for elaborate physical models and actual tests. A substantial body of knowledge verifying the physical laws and principles is a prerequisite to any modeling process, and ultimately a full-scale test is required to verify model results, but simulation is a powerful tool for predicting performance.

Complicated though physical systems may be, they are infinitely more simple than social systems, where responses can be random, nonlinear, and unpredictable. Despite these limitations, modeling has the ability to provide decision makers with a much more comprehensive review of available options, and Chapter 7 discusses some of the physical, financial, and social models available in this still embryonic application to local government activities.

In our present state of evolution in planning and forecasting, we must know where we have been in order to understand where we are going. Chapters 8, 9, and 10—on the environment, energy, and the revolution in electronics—are written to provide the reader with a sense of how these driving forces have arisen and how they are influencing the decisions we are making. They are surely not the only driving forces affecting local governments as we move through the 1980s, and they may not be the dominant ones.

Certainly, the question of financing local governments is a major issue as the rosy glow of Great Society initiatives grows dim in the struggle to make slow-growing revenue sources sustain inflation-fueled budgets. This budget gap has been further strained by enormous growth in demand for human services that are complex, difficult to provide, and taxing of the physical and financial resources of the communities that seem to be pushed ever more deeply into them. However, these issues are less technology oriented and better left to another book whose focus is not technology.

Chapter 8 traces the environmental legislation that was initiated in the 1950s and 1960s and blossomed in the 1970s. In environmental terms, the 1970s were years of major accomplishment. A sophisticated lobby and infrastructure of people dedicated to protecting our environment evolved, and national legislation in land, air, water, and other areas developed. The gains of the 1970s were hard fought, and a substantial number of private citizens and public servants are committed to the cause of protecting those gains. In the 1980s, we find ourselves under severe economic pressures induced at least in part by the ascendance of the Organization of Petroleum Exporting Countries

(OPEC) and its cartel activities in world oil markets. There is no doubt that environmental benefits come at some cost, and there is rising pressure to trade those benefits off in favor of energy and economic considerations. However, the environmental legislation is on the books, its protectors are in place, and, despite the fact some accommodations will be made, there will be no substantial retreat from the gains already achieved. Existing environmental regulations will continue to shape what localities can do in the 1980s.

Chapter 9 emphasizes that the era of cheap energy is over. Abundant supplies and government regulation ensured that energy was a cheap commodity until the 1970s. The combination of ready supplies and federal price controls caused a large-scale shift away from coal and into oil and natural gas as primary energy supplies in the middle of this century, just as we had shifted from wood to coal in the previous century. Those same forces that ensured cheap, easy-to-handle energy in the form of oil and natural gas forced us to go overseas to maintain its abundance. We gradually shifted our consumption patterns through the 1950s and 1960s until, by the early 1970s, we were importing almost half the oil we needed for domestic consumption. Coal consumption had dropped and atomic power got mired in regulations and construction and safety problems, thus making us more and more dependent on oil and natural gas for energy. The stage was set for cartel action, and OPEC seized its opportunity in 1973.

In the ensuing years, we have seen a steady rise in oil prices and a tightening of our economy as fuel costs made themselves felt everywhere. Although the changes to our economy have not been as drastic as some predicted, they have been pervasive and we are still undergoing our transition. Such major industries as automobiles and building construction are affected, and, after a long period of conflicting signals, the message is clearly arriving: Energy costs will continue to rise, and our ability to function locally, nationally, and in the world community will be paced by our ability to develop alternatives to imported oil.

Energy problems and environmental controls will continue to constrain us. By contrast, the revolution in electronics will create new dimensions and boundaries. Our energy production and utilization methods are mature. Changes brought on by energy prices will be far reaching, but they will be evolutionary rather than revolutionary in nature. The same can be said of our transportation modes—rail, ship, motor vehicle, and aircraft. All are highly evolved, and major changes to our transport system have projected time spans that reflect its advanced state of evolution. In air transport, the youngest industry, the technology that will carry us through the end of this century is either in place or on the drawing boards.

Electronics, by contrast, is still sitting on the frontier. The transistor, the device that ushered in the latest revolution, is less than 30 years old. The transistor made possible the large-scale mainframe computer and its smaller-scale derivative, the minicomputer. The transistor was also the building block for integrated circuits that have evolved into the microprocessor—the computer on a chip. All these devices provide the ability to command, control, and communicate at less cost than ever before, and costs are still falling as capability rises. As chapter 10 indicates, the electronics revolution will impact the home, the factory, and the office, and its impact will increase with time. Any book that discusses technology must include this revolution as an integral part of the discussion.

To be sure, there will be some clash between this emerging technology and our ability to cope with it. Around-the-world teleconferences that convey sight, sound, and intonation may be possible today, but it will be quite a while before electronic communications can provide the intuitive measurements that come from face-to-face meetings and are considered so essential to the traditional decision-making process. Similar conflicts may occur in industrial electronics use, where automation has the potential to change the mix of skills, elevating some and eliminating others. Labor organizations and people whose skills will be obsoleted will fight automation.

Data processing may face a similar backlash—a fight against the big brother syndrome engendered by the computer's potential to aggregate and display vast amounts of data about any person.

Nonetheless, the potential benefits from the revolution are substantial, and they will continue to fuel the drive for technology for some time to come. Our society will be tested by its ability to harness the technology and put it to our advantage.

The last chapter in this book, "Putting It All Together," discusses technology assessment within a locl government and gives an example of the application of the process to local government situations. Technology assessment is in itself a new discipline and requires some special skill to implement. The chapter outlines two approaches to assessment: It can be established as a separate discipline within an organization, or it can be integrated into another decision-making process, such as the establishment of the budget. In any event, it is a soft discipline. Its output is advisory, and it will not be terribly easy to tell the difference between good results and a mediocre effort.

Because of its nature, an assessment process must be carefully structured from the outset, and considerable care must be taken to ensure that the people selected to perform assessments have the background and skills necessary to conduct an assessment, as well as the proper attitude and motivation. A poorly organized and staffed assessment team is worse than none at all: In the long run, it will give a false sense of security and waste money.

Chapter 11 gives an actual example of the assessment process applied to the selection of a telephone system for a local government. Telephones as electronic systems are undergoing rapid technology changes, and these changes are occurring at the same time as regulatory changes concerning the provision of telephone services. The compound of technology and regulatory changes produces a complex situation that is particularly amenable to the assessment process.

There is another side to the assessment process as well. It can identify some of the organizational issues that must be worked on to assure successful implementation. The last section of Chapter 11 outlines how some of those issues were addressed to ensure that the selected telephone technology would fit into the organizational setting of a particular local government.

CHAPTER 2

Technology and Society

Looking through the glass that isolates the carefully climate controlled interior from the elements outside, a viewer sees miles of concrete ribbon strung with power and communication lines. Automobiles snake out along one of the concrete corridors leading to a major airport, where outsize jet planes take off in the distance. A gentle breeze carries away the smog and haze that usually envelop the city, causing the silver and white planes to stand out in stark contrast with the pale blue sky. Almost immediately below the building, the central terminal of the rapid rail system appears along with the neon trademark of the world's largest soft-drink bottler. In another direction, hidden from view by industrial plants, a wastewater treatment plant processes the city's sewage and returns it to the same river where seven miles upstream water has been removed and processed through the water supply plant for the city's use. A little farther up the river lies a massive electrical generating plant that supplies most of the power for the metropolitan area. Off to the left, a queue of cars is forming along a major connecting street. A new computer-controlled streetlight system has just been installed and is still being debugged. The glut of traffic stymies the refuse collection trucks and the buses as they follow their daily routes. The horse-mounted police officer directing traffic strikes an incongruous note in an otherwise mechanized scene. Inside the building, the viewer's attention is distracted by the clatter of a printer attached to the communications terminal and the simultaneous ringing of the telephone.

A TECHNOLOGICAL SOCIETY

Ours is a technological society: highways, climate-controlled buildings, telephones, and transcontinental air transportation are products of the age. So too are congestion, smog, and noise. Technology has changed the face of our land and rapidly reshaped the way we live. We, as an industrial society, have a pressing need to understand both the positive and negative aspects of technology and its impact on our lives so that we can maximize the intended benefits and minimize the harmful side effects. This book will examine some of the tools that can help us predict the consequences of technology choices, particularly as they relate to the many local governments around the country.

The twentieth century is dotted with examples of industrialization and its side effects. Technology and mechanization have revolutionized food production. Through the use of intensive cultivation, fertilizers, and mechanized equipment, the United States has become a primary producer of the world's food supply. With only 15 percent of the world's agricultural land, the United States provides the majority of the world's surplus grain. Our productivity is significantly higher as well. This intensive cultivation has brought with it a transportation and distribution infrastructure that makes it possible to bring food rapidly to distant cities and foreign countries.

It has also brought some harmful side effects. DDT, a common pesticide, accumulates in the bodies of animals exposed to it and makes its way up the food chain into the meats we eat, causing damage all along the way. Because of its toxicity and its tendency to accumulate in living tissue, DDT has been banned for widespread agricultural use.

Kepone, another widely used pesticide with similar characteristics, has contaminated the entire lower reaches of the scenic James River in Virginia and can possibly endanger the Chesapeake Bay, the James River's entry to the Atlantic Ocean.

This pesticide was produced for several years in a converted filling station, with little regard for its toxicity and virtually no safety precautions. The workers who were exposed to the Kepone suffered permanent damage, and the residue from the manufacturing operations trickled through the local municipal wastewater plant, ultimately making its way into the James River.

This situation was uncovered when the municipal treatment plant's biological processes were interrupted by a large slug of Kepone waste. Subsequent investigation revealed that Kepone had settled into the sediment of the James River and was being absorbed into the tissues of fish in the river. The Kepone plant has been closed down, but the James River may be off limits for fishing for decades, and the plant workers and their offspring will be examined for years before the total human damage from the Kepone can be assessed.

Agricultural fertilizers also are not without penalties. Although they make intensive cultivation possible, their production requires significant quantities of energy and petroleum feedstocks. Thus, agricultural production increases total fossil fuel consumption, further adding to world energy consumption. Fertilizers as nutrients contribute to the pollution of rivers and streams as well. In suburban and rural areas, they make rivers and streams work harder to neutralize the fertilizer runoff and stimulate unwanted plant growth that can spoil recreation areas and change the food chain in the water bodies.

We are living in a chemical age. Pesticides and fertilizers are chemical combinations of basic elements. So are plastics, synthetic oils, nylon and polymer fabrics, paints, perfumes, adhesives, and food additives. Synthetic materials are replacing natural materials at an ever increasing rate in everything from food additives to construction materials. We are just beginning to understand what these chemicals take from us as we rush headlong into the benefits without weighing the costs. We have placed strict limitations on the use and disposal of mercury, arsenic, and heavy metals. Beryllium, a space-age wonder material, poisons its fabricators. Benzene, a component of gasoline, is suspected of causing cancer. Asbestos, our time-tested insulating material, has also become a villain. Yet we will continue to devise more complex formulations with even more potential to interfere with the delicate processes that support living things.

THE AUTOMOBILE

A major technological accomplishment of the twentieth century, the automobile, has brought dramatic changes to our society. Mobility has increased; suburbs connected by highways have sprung up around cities; vacation lands have become accessible to all; and the ability to span long distances has produced substantial cross-migration, melding regional differences and weaving a single cultural fabric through the country.

Mobility requires low-cost transportation, and mass production techniques have lowered the price of automobiles so that they are available to almost everyone. In the process, a major industry has been created. Automobile manufacturers employ millions of people. Basic industries that support automobile production—iron, steel and other metals, glass, rubber, plastics and fabrics, electrical components, mechanical components—have burgeoned with the auto industry. The automobile has also spawned a major service industry. Gasoline stations, service centers, and tire and accessory stores are important parts of our economy. All in all, motor vehicle production and operation accounts for a substantial percent of our gross national product. Economically and socially, motor vehicles have made a major contribution to our society: They have produced unprecedented personal mobility and freedom; they have converted the country from isolated regions to a unified whole; they have created new industry, new jobs, and a larger economy. Unfortunately, they have also caused congestion, pollution, urban sprawl, and enormous energy consumption.

No one could have predicted the impact of the automobile on our cities and towns. Some of the older cities in the United States matured before the ascendance of the automobile. These densely populated cities, particularly along the East Coast and in the Midwest, were not designed for the traffic flow they now accommodate. Even some of the later-blooming cities in the Southeast and the Southwest are barely able to handle auto traffic. Highways and bypasses, originally constructed to aid traffic flow, seem only to encourage more traffic. Often the end result is vehicular traffic that moves more slowly than pedestrian traffic, an ironic by-product of the automobile age.

The auto has caused another unpleasant side effect: the debilitation of mass transit systems in our major cities. Mass transit systems, which were built to move people efficiently in the first half of the century, declined in ridership as the automobile became a dominant mode of transportation. The declining ridership and decreased revenues debilitated the transit systems. And as the transit systems decayed, more riders shifted toward automobiles, further adding to road congestion.

The automobile was responsible for the growth of the suburbs by making them accessible to millions of people. Easy, rapid access to outlying areas has made it possible to mix the economic life and pace of the cities with the quiet of the suburbs and rural areas. The suburbs, however, have grown at the expense of the cities; suburban and rural populations have increased as city populations have decreased. Mobile, affluent citizens have fled to the suburbs, leaving the cities with smaller tax bases to support the poor, the indigent, and the elderly. However, the automobile has not produced unlimited suburban prosperity. While many suburbs have fared well, some have been burdened with uncontrolled population growth, strip zoning, excessive traffic, and high taxes to support the rapid development of facilities and services.

In the cities and the suburbs, the concentration of motor vehicle traffic led to another unanticipated problem—air pollution. Engines that power automobiles burn gasoline in a combustion process, taking in oxygen from the air and mixing it with hydrogen and carbon contained in the gasoline to provide heat energy, which is converted to mechanical energy. The products of combustion—carbon monoxide, carbon dioxide, particles, and unburned hydrocarbons—are emitted to the air through exhaust pipes. Some of the nitrogen present in the air with the oxygen does not contribute to the combustion process but is changed to oxides of nitrogen inside the engine. The combustion products are damaging: Carbon monoxide is toxic; carbon dioxide dilutes the air; unburned hydrocarbons and nitrogen oxides are toxic and react photochemically in the presence of sunlight to produce smog; and particulate in large quantities blocks out sunlight, damages the lungs, and layers dust on buildings, people, and other objects. Although other combustion sources produce the same pollutants, the automobile is the No. 1 contributor to pollution in urban areas.

The automobile is a principal contributor to our present energy predicament as well. Since the turn of the century, growth in automobiles and growth in oil consumption have paralleled one another. Each year the number of motor vehicles and the average mileage driven per vehicle have increased, and currently gasoline for motor vehicles accounts for about a third of the 17-odd million barrels of petroleum consumed per day in the United States. In the 1950s, faced with growing oil consumption and leveling production, we turned overseas for additional supplies of oil, thereby setting the stage for the energy crisis we are now facing. Gasoline production consumes the equivalent of more than three-quarters of the oil we import today.

In 50 years, the automobile has grown from a curiosity to an indispensable part of our society. But in so doing, it has caused dislocations and disruptions that we were ill equipped to perceive and quantify. Clearly, the impacts of such a far-reaching technology need to be perceived in advance, rather than in retrospect, if we are to do a better job of avoiding undesirable consequences in the future.

FORECASTING AND ASSESSMENT

Where are we going? What does all this mean? The issues raised at the end of the previous section lead to the fundamental point of this book: If we are going to live in a complex technological society, then we must develop some means to shape the technology so that we can minimize its negative impact and maximize its intended benefit.

As a society and as individuals, we are constantly making choices. And as we develop greater influence over nature—as technology grows—we have the potential to alter so-called natural processes on a grander scale. However, as we make larger changes, we also create larger consequences, and the choices become more critical. Technology forecasting and technology assessment are two elements in an array of tools that can be used to improve our decision-making processes and our understanding of the consequences. *Webster's Seventh New Collegiate Dictionary* defines *technology* as "a technical method of achieving a practical purpose" and "the totality of the means employed to provide objects necessary for human sustenance and comfort." Technology, therefore, can be a product, a process, or even an organized body of thought, and this book will treat technology in its broadest context.

Technology forecasting is an art that formalizes the decision-making process by allowing us to systematically (if imperfectly) measure the possibilities for success in a specified technological course of action, given some prior knowledge of previous actions and the ability to choose possible alternative actions. Technology forecasting by definition is future oriented. Technology assessment, on the other hand, focuses on the consequences of a particular course of action, rather than on the action itself. Technology assessment can be retrospective, looking at the results of a previous course of action, or it can be near- or far-term future oriented, projecting the consequences of actions already taken or possible future courses of action. In this book, technology assessment will be defined as measuring and anticipating the near- and long-term consequences of new technology.

There are elements of assessment in any technology forecast, just as there are elements of forecast in any technology assessment. Although the two are not totally separable, it may be convenient to think of technology forecasting as focusing primarily on the technology and technological decisions themselves, while technology assessment measures the economic, political, and social as well as the technical consequences.

Antecedents

Technology forecasting is derived from our need to provide a strong national defense, and it was originally established as an adjunct to military decision making. As our level of sophistication in weaponry grew, so did the cost and time required to produce the machinery for war. The options were greater, and so were the choices for our adversaries. What weaponry do we have now? What will we have in the future? What should we have in the future? How does it compare with what our opponents have? What are their military situations likely to be in the near and far future? What will be our investment to maintain parity of force? Can we maintain parity with present and projected resources? All these questions needed to be answered in a methodological rather than an intuitive fashion, and thus rose technology forecasting as a formal art in think tanks, among military contractors, and in the armed services themselves.

That is not to say that technology forecasting was invented in the 1950s and 1960s. Although it matured as a discipline during that period, people have always been forecasters. In the 1920s when the National Advisory Commission on Aeronautics (NACA, the forerunner of NASA) was formed, some critical decisions were made about the type and level of research investments that would be made in the then emerging science of aerodynamics. Those decisions could not have paid off more handsomely, since they set the stage for our present-day preeminence in the design and manufacture of commercial jet aircraft. The aerospace industry today is a major producer of new technology, a substantial employer, and one of the manufacturing industries where our exports far exceed our imports. The industry absorbs skilled people and makes a positive contribution to our balance of trade as well as our national economy.

However, technology forecasting, of itself, is not enough. If the assumption is made that no one has unlimited resources (a particularly valid view at this time), then a decision to advocate a particular technology must be accompanied by information about investment and consequences. Technology assessment, which measures consequences, thus grew as a corollary skill to technology forecasting. It answers the question, "If we go there, what does that mean?" and has only lately been recognized as a discipline in its own right. Technology forecasting has a place in civilian as well as military federal affairs and should have a place in state government as well. However, as we move further away from global considerations and closer to individual citizen needs, measuring consequences becomes more critical than strategic planning and assessment takes precedence over forecasting.

Pitfalls of Forecasting

There is always concern about the accuracy of forecasting and assessment. Our ability to predict and measure is predicated on our knowledge of the past and on some feeling for what, for lack of a better word, will be termed a "reasonable" pace and direction of future happenings. This all assumes, of course, that the activities we are concerned with flow from one another and can be plotted on some kind of curve. Given this assumption, the task of forecasting and assessing is to find our present place along the continuum and predict whether the curve will go straight ahead, or up, or down and at what rate.

All this continuous activity ignores what plagues forecasters— the breakthrough. A breakthrough is an advancement of knowledge or the development of a device that alters and accelerates the pattern of a particular technology, creating a discontinuity in an otherwise smooth projection curve. The jet engine was a breakthrough in air transport; the discovery of the transistor ushered in the electronic age, a revolution that is still unfolding. The problem with breakthroughs is that they cannot be predicted. You can see them coming, but you can't tell exactly when they will happen and what they will mean. If they change the course of a technology, the consequences are difficult to assess. Forecasters are held hostage to breakthroughs: If they predict a small change in the course of future events, they are accused of lack of courage; if they predict a revolutionary change that proves to be evolutionary, then they are called foolhardy.

All this may lead to the question: Why bother to forecast or assess at all? This can be easily answered by two basic tenets of decision making. First, if there are two possible courses of action, then there is a 50 percent chance that an arbitrary choice will be correct. If some supporting information is available to assist in the decision making, the probability of making a correct decision increases. Second, very few decisions are single point and irrevocable. There is usually a series of decisions required to pursue a course of action, and an initial wrong decision can be modified by subsequent downstream decisions.

THE DECISION-MAKING PROCESS

Decision making for the future, correctly implemented, involves a sequence of activities repeated over and over in time until an action reaches its conclusion. The steps are: decide, plan, implement, and evaluate. (See Figure 1.)

The first step, deciding, starts with one or more possible courses of action. A forecast is made of the alternatives. How far can you go? Are all the elements for success available? How long will it take? What obstacles need to be overcome? An assessment of the consequences follows right on the heels of the forecast. Will it accomplish its intended function? What changes will it cause in the organization? In the outside world? Will it be accepted? What

FIGURE 1. Sequence of activities in the decision-making cycle.

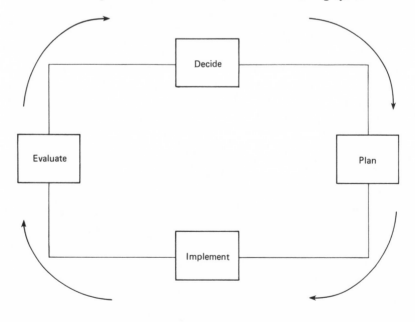

good will come from it? What are the consequences of failure? How strong are they? This forecast–assessment process leads to a decision that sets in motion the planning process.

Planning is the process of making way for the change, lining up the resources and elements necessary to make the change, and charting and relating activities. It ties together the detail elements of the decision into a workable scheme.

Implementation is the process that turns the planning into reality. Significant resources—time, money, and materials—are committed to the task during the process to ensure completion. Coordination is required to ensure that all the elements of the activity are present at the right place and time and that each of the elements complements rather than conflicts with each of the others.

Evaluation is the final sequence in the process. The evaluation can be a current or retrospective assessment. On the one hand, it can measure actual results against projected results. If the activity is not proceeding according to plan, it provides the feedback necessary for midcourse decision making and corrective action. On the other hand, if the activity has been successfully completed, it provides an assessment of the activity itself and its impact on the larger world for future similar situations. If the activity has been a failure, careful assessment can provide benchmarks for future successes.

Decision making is thus a continuous process. It may take place in a step fashion, but if properly done, it forms a continuous cycle and the assessment process is given tremendous strength. As any activity matures, the focus narrows and the boundaries become more clearly delineated, allowing us to successfully narrow the forecast–assessment process to ensure a succession of informed decisions leading to the ultimate goal.

THE IMPORTANCE OF FORECASTING IN OUR SOCIAL SYSTEM

One need only look at the index of leading stocks through the decades of the twentieth century to understand that our society is constantly evolving. Stock market activity reflects the strength of mature companies through the trading of so-called blue chip stocks. It also highlights the up-and-coming growth or glamour industries through the trading of speculative stocks. In the early 1920s, basic food producers formed the backbone of our commerce and finance activities and the stock market reflected that trend. In the 1940s and 1950s, manufacturing companies held sway, while in the 1960s and 1970s, electronics made the major impact.

Industries must stay with their marketplace, shifting with changes in taste or demand in order to stay healthy. Companies, therefore, are required to continuously forecast the expansion of new markets as well as the phasedown of existing markets and to plan their product lines accordingly. The recent plight of the American automobile companies is an interesting example of failure to follow the market. The automakers found themselves in a severe sales and financial crunch when sales of mid- and large-size automobiles dropped precipitously through 1979 and 1980.

It can be argued that the change in consumer taste was coming and that the companies should have been able to forecast the shift. It can also be argued that the change was a forecaster's bane—an unpredictable discontinuity in the sales curve, since the shift occurred in three to six months, whereas the production lines require three to five years to readapt to different vehicles. In any event, it is a little too early to understand how shift in consumer taste was miscalculated. However, it may be too late for at least one auto manufacturer, which has been forced to go to the government for loan guarantees to support its staggering financial losses.

Government, unlike industry, is established strictly to provide order and governance in the conduct of citizens' affairs and is specifically chartered to remain in place indefinitely. (It may be argued that there are too many levels of government and too many obsolete government entities, but that is not the subject of this book.) Although governments remain in perpetuity, their goals and functions shift as our society evolves. The United States' growth in relation to other nations from the time of its inception gives ample evidence of changing priorities. In the eighteenth and early nineteenth centuries, the nation concentrated on keeping its cultural ties with Great Britain, but established its own identity as a nation and focused on growth and consolidation. The mid-nineteenth century focused still on internal questions— the legitimacy of secession and expansion of our boundaries in the West. A shift from isolationism occurred in the twentieth century with military activity in Cuba and two world wars to end all wars. The second half of this century has seen the end of all the simple answers to world politics, and we have learned that the United States does not hold all the cards in the game, and our intentions and motives have been questioned and ridiculed in world forums and among ourselves. If our external dimensions have changed, our internal dimensions have changed even more.

Our evolution at a national level makes a clear case for a constantly changing role for government. And as technology has increased our capacity for communications, transportation, trade, and the utilization of world resources, it has intensified our relationships with other governments. We cannot make decisions in isolation, and the elements of future change become more complex as our interrelations become more complex.

As our external relationships become more complex, so do our internal relationships. Just as there is a need to plan for and anticipate change at the federal level, there also exists a need at the state and local levels. Each level, however, anticipates change in accordance with its role in the governing process. Federal government is the highest level in our political structure. The constitution provides for the control of all regional and national functions and delegates functions within the state to the states themselves. In recent years, these boundaries have been blurred, and the flow of federal revenues to states and localities has considerably extended federal influence, but that discussion is reserved for Chapter 3.

Major areas in the federal purview are the economy, national defense, and citizen health and welfare. There is a strong need in each of these areas to forecast where we are going and to take appropriate steps to ensure that we move toward the attainment of our larger social goals with a minimum of side effects. Two of our present major issues, energy and environment, are heavily technology dependent; they will be addressed later in this book. Overall, at the federal level, there is a need for strategic long-range forecasting, as well as a thorough ongoing assessment of federal initiatives, which by their nature affect a broad spectrum of people.

States have responsibility for such issues as transportation and water supply within their boundaries. These activities are often handled in conjunction with the federal government but by definition are the provinces of the states. States must also be in a position to cope with issues of population growth and shift, and are responsible for husbanding state natural and economic resources for the general good of their citizens. They also exercise licensing and regulatory functions, as for instance, the regulation of power and communications utilities and the licensing and regulation of motor vehicles.

States have the right, as does the federal government, to tax citizens for services. In general, the tax base of most states provides enough flexibility and resources to allow them to discharge their obligations to their citizens. The flexibility of state taxing powers provides states with much independence from the federal government, and this is a key issue in state–federal relationships. Because of their involvement with resources, taxes, regulation, and control, states have a need to forecast and assess, but on a more restricted scale than at the federal level.

At a local level, the emphasis shifts largely from strategic planning to meeting more immediate needs. Local governments provide the first level of services to citizens and serve as the vehicle by which citizen needs, as well as state and federal mandates, are met. This complex service delivery role emphasizes response rather than elaborate planning, and it is at this level that technology assessment can have its greatest impact on our structure of governance.

CHAPTER 3

Technology Assessment at the Local Level

In this chapter, we will look at what a local government is, what it does, and the place of local governments in our overall economy. We will also look at some barriers that build in a bias against an elaborate planning and budgeting process, and conclude with a summary of the importance of technology assessment in the local government process.

WHAT IS A LOCAL GOVERNMENT?

Local government is defined by the Oxford English Dictionary as "the administration of a town (or other limited area) by its inhabitants as distinguished from such administration by the state at large." General-purpose local governments—those units chartered to provide broad administrative services to citizens—and other service entities that meet specific specialized needs, such as service districts, school districts, and planning areas, can be looked upon as not-for-profit public service corporations chartered to meet the needs of everyday communal life. Although these local governments are variously described as cities, counties, towns, boroughs, villages, and parishes, the two most widely used designations are city and county.

Counties and cities trace their origins to our early English tradition. In the United States, counties are the level of government lower than the state, and all states are divided into counties or county equivalents, such as parishes or districts. Counties vary widely in size and function. In New York and other mid-Atlantic states, counties share their responsibilities with a crazy quilt of cities, towns, villages, school districts, and special service districts that, more often than not, have overlapping boundaries. In Virginia, the major urban counties look and act like cities. Although there are historic exceptions, cities and counties in Virginia are parallel units of general government and provide all essential services from police through schools within their geographic boundaries with little or no overlap in services. In contrast with Virginia, however, most states have cities that are municipal corporations with separate governing structures established within their counties.

Cities were established originally as centers of commerce and industry and have traditionally provided the greatest level of services. This overlap of jurisdiction has produced some duplication of services, particularly as counties have moved from the status of rural government entities to deliverers of services to major suburban populations. Although cities like New York and Chicago dwarf most counties, units such as Nassau County in New York and Los Angeles County in California have substantial populations.

The city–county situation is further complicated by the question of annexation—the addition of adjacent land to enlarge municipal boundaries. In Virginia, there is a 10-year moratorium on annexation of unincorporated land by cities, while in Texas and Oklahoma, the cities pursue an aggressive annexation policy as counties provide services to a diminishing rural population and a diminishing land base.

Although there is significant variation in population, powers, and responsibilities, all local governments are characterized by a structure whereby citizens have assigned the responsibility for governance to locally elected officials who provide direction to an organization that provides basic services to its citizens.

A SERVICE DELIVERY CORPORATION

As a service delivery corporation, the local government provides for public order, life and property protection, regulatory functions, and human services. It also provides such basic physical services as transportation, water treatment, wastewater disposal, solid waste collection and disposal, and sometimes electrical power generation and distribution. To pay for these services, it levies and collects taxes.

Public order is provided by the legislative body through a system of ordinances that complement federal and state law. Protection of life and property is provided through the public safety system—police, fire, and sometimes emergency medical services. The criminal justice system provides the link between public order and the police enforcement section of the public safety operations. The justice system provides for prosecution of violators of laws, for judgment of innocence or guilt, and for punishment in both criminal and civil cases.

Localities perform a major regulatory function by providing and administering zoning regulations during the conceptual land development process. They also regulate the structures that are placed on land, by the building code and inspection process. In areas that involve public health or sanitation, there are provisions for inspection and regulation. Commercial and business licenses are another vehicle for regulation in the public interest.

Local governments also provide people-related services: human services, such as hospitals, medical and mental health facilities, clinics, and nursing homes; cultural facilities, such as libraries, sports arenas, civic centers, and concert halls; and recreational facilities, such as parks, zoos, and recreation areas.

The single most costly service, education, must be put in a category by itself. Education has a substantial place in our society, and its impact can be clearly measured at the local level, where the public school system, the backbone of our primary and secondary education system, is financed and operated. Primary and secondary public schools usually account for 50 percent or more of a local government budget. The school network is substantial, and the sheer size of our public school facilities is testimony to the importance we place on education. In many suburban communities, the school district is probably the largest single occupier of building space. Education is also labor intensive. State mandates limit the teacher–pupil ratio, and when supervisory and support personnel are added, the ratio of school system employees to school-age children approaches 1 to 20 or fewer in most public school systems.

Local governments construct and maintain streets and roads for vehicular and pedestrian traffic, and in many instances provide public transportation to facilitate rapid movement of people within and across their boundary lines. They also provide water for residential, commercial, and industrial consumption by providing water treatment plants and water distribution systems. In addition, wastewater collection networks and treatment plants are provided to handle the sanitary and industrial waste products. The solid waste leftovers from consumption—garbage, refuse, litter, and trash—are collected and disposed of as well. More than a few municipalities are owners of electrical power plants, or retailers and distributors of purchased electrical power, and operate elaborate electrical distribution systems.

The services and facilities listed above represent the expenditure side of the picture. The income-producing side includes valuing property and levying taxes, collecting fees for services rendered, and administering a wide variety of licensing and permitting operations. Increasingly, local governments are collecting sales taxes authorized by state legislatures.

This large revenue-collecting service delivery organization also needs a corporate staff and senior management to implement policies and ordinances; a legal staff to defend the jurisdiction, draft local ordinances, and follow federal and state laws; a purchasing staff to make acquisitions of real estate, goods, and services; a personnel department; a maintenance staff to keep the physical machine well oiled; and, of course, a public affairs office to keep the public informed of its activities.

TECHNOLOGY AND SERVICE DELIVERY

Technology has a strong role to play in the delivery of facility-based services. Transportation—air transit terminals, rail and bus transit equipment and facilities—bears the markings of the high-technology air and surface transit industries. Streets and roads can benefit from new tools in the design and analysis process, and road construction requires major investments for mobile construction equipment. Water treatment and distribution involves huge quantities of water and a companion investment in physical facilities, including mechanical and electrical machinery, instrumentation, and power control and distribution apparatus. Wastewater treatment facilities have received increasing attention and significant federal funding as we realize that our oceans, rivers, lakes, and streams cannot take an endless stream of pollutants. As our environmental regulations tighten, the wastewater treatment plants become more sophisticated, and simple treatment processes are giving way to technology-based systems with substantial process controls.

Solid waste collection requires an investment in sophisticated packer trucks. Solid waste disposal is coming out of a long period of neglect, as open dumps become sanitary landfills that meet strict environmental design and operational requirements and as resource recovery plants that take materials or energy back from the waste stream become more practical. The resource recovery facilities have much more in common with industrial process plants than with the landfills they are supplanting.

Police and fire agencies, always equipment-based organizations, are moving more heavily into communications and data processing as they seek to plan and manage while maximizing human resources.

The technology orientation of the physical service delivery efforts is easy to understand. Human service delivery, however, imposes less obvious but substantial demands for technology. Each service delivery system has its own needs for shelter, communications, and transportation. Elaborate systems are needed to catalog and track recipients of assistance, and efficiency and effectiveness must be measured. Questions such as "Are we providing the service people need at a cost that reflects the value of the services?" have to be asked and answered constantly. Overall, there is a need to provide coordination and management to this corporation so that it can manage the present, plan for the future, and respond to requirements from its citizens as well as to changing state and federal requirements.

BARRIERS TO INNOVATION

The nature of local governments and the boundary conditions created from above by other tiers of government as well as from below by citizens create circumstances where innovations—new approaches and technical changes—are difficult to carry out. Any innovation—whether it be a hardware item, such as a new wastewater treatment process, or a soft technology, such as technology assessment—must be carefully nurtured with due care for some of the institutional barriers that can prevent its acceptance.

Major barriers include the focus on land use, pressures from above and below, the two-year election cycle, organizational structure, and the fish bowl environment.

Land Use

In suburban communities, residential property owners have a major investment in their dwellings and look to them as a source of psychological as well as physical shelter. Planning and zoning issues are of major concern to property owners, and the question of property development can be a drawn-out, time-consuming issue. Other government activities are directly related to maintaining the character of the community or the quality of life as well. Although taxes range from moderate to prohibitively high across the country, individual income is high enough to sustain the emphasis on maintaining the character of suburbs while sidestepping the question of more intensive land use that is often brought on by a declining tax base.

In urban areas, the emphasis is on retention of housing stock, issues of density, and how to maintain the character of a particular locality, while maintaining the continued investment necessary to provide growth in a jurisdiction's tax base. The larger, more mature cities, in particular, have expanded into areas of governance far beyond what is provided by the suburbs. Urban development and renewal, mass transit, airports, harbors, and extensive human service programs have been added to the standard array of physical facilities and citizen services that cities traditionally provide. Although these additional facilities and services diversify the city's role, they ultimately wind up as land-use related, since a city's ability to sustain them relates back to its tax base, which in turn relates to the value of real estate and the taxes that can be derived from the physical structures, businesses, and individuals who populate the land.

The whole question of land use is a major issue in any city or county and deserves a major time investment by the elected and appointed leaders of a jurisdiction and their support staffs. Although there are elements of technology associated with land use, they are peripheral and not really central to the social decision-making process, which receives so much attention from local governments.

Pressures from Above and Below

Federal activities are usually broad in nature and focus on providing services at a national level through lower tiers of government. At a state level, some more direct provision of services occurs, but most services are aimed at all the state's regions and citizens, and much of the responsibility for service delivery is again passed through to localities. At a local level, the first level of contact with citizens, the demands become more immediate and the activities center around operational service delivery.

In many areas, local governments respond to state and federal dictates rather than initiating activities. This passive response is due primarily to city–county–state relationships and the all-encompassing nature of federal regulation and funding.

By definition, local governments are creatures of the state and may exercise only those powers specifically assigned by the state. This downward flow of power makes local governments subject to the state legislatures for charter amendments to provide adjustments to their powers. Nowhere is this more greatly felt than in the area of tax growth, where the relatively inflexible real property tax dominates, and localities must petition the states for more flexible taxing powers, such as sales and income taxes. Local governments are also subject to new initiatives from state legislatures. Although there is increasing resistance to performing these new activities without receiving the resources to accomplish them, the downward flow of power still creates a situation where local governments must respond to state initiatives.

Federal powers, although in theory less direct, are in actuality more all-encompassing than state powers, and they flow fundamentally from the ability to force compliance with federal policies, guidelines, and activities in exchange for federal revenue. The constitution provides the federal government with powers that are complementary with state authority. For instance, the federal government has purview in interstate issues, such as commerce, communications, and utilities, and in constitutional issues, such as discrimination. In theory, the federal laws are operable only in areas directly affecting multistate regions, the nation as a whole, or classes of people. In practice, there has been a significant extension of law and policy in local activities by making the implementation of federal policies a condition of receiving federal revenues. And with the growth in federal monies in local budgets has come a corresponding growth in federal influence.

Examples of federal influence abound—some valid, some not so rational. For instance, the Davis-Bacon Act requires construction contracts that receive full or partial federal funding to pay workers the prevailing local wage rate as determined by the Department of Labor. Interstate Highway 66 in northern Virginia was to have shared right of way with the Washington Metro Rapid Rail

System, and the provisions of the Davis-Bacon Act were construed to mean that workers constructing the highway were to be paid the same wage as rapid rail construction workers. This was about double the prevailing wage paid to road workers in the state. The decision would have drastically increased the cost of the highway, distorting road construction wages in the process. This decision was fought by two successive Virginia governors until it was finally brought to a successful conclusion in 1979, with the ruling that highway workers be paid prevailing wages for similar work in the area.

The Environmental Protection Agency (EPA), which provides up to 75 percent of the funds for the design and construction of wastewater treatment plants, uses its financial leverage to enforce water quality requirements locally, regionally, and nationally.

More recently, there is a requirement to make small and minority business an integral part of any federally supported activity. Major contract and construction activities require minority participation programs, and each federal department and agency is required to set aside contracts for materials, construction, and services specifically for performance by small and minority businesses.

All these activities constitute requirements that local governments must meet in order to assure a continuous flow of federal revenues. Federal funding is also used to provide incentives for such things as downtown revitalization, the construction of environmental facilities, and the development of airports.

There are clear signs, however, that the federal government is also willing to use its funding activities in one area to ensure compliance in other areas. This is particularly true in areas that have broad national support but significant local resistance, such as integration and air quality. The government has turned increasingly to cutoffs of federal funds in all its activities in cases where state and local governments have obviously been dragging their feet in the implementation of federal mandates.

While local governments are coping with state and federal mandates, they must still serve as the vehicle by which such basic community services as refuse collection, fire and police protection, emergency medical services, and water supply are performed. These basic services are immediate and vital to life and health, and citizens have come to expect them to be delivered continuously or on a scheduled basis with a minimum of interruption. Because of the vital nature of the services they deliver, and because of their proximity to the citizens they serve, local governments by nature must be operations oriented, with quick response as the mode of operation. Individual citizens have difficulty communicating with other layers of government and often perceive the local government as the only level they can influence. Localities therefore stress local control and sensitivity to citizens' needs and cannot afford the kind of orderly, rigid planning process that becomes more the rule as a government becomes separated from its constituency.

The Two-Year Election Cycle

The customary two-year election cycle for local government officials and their vulnerability to small local issues create an environment where immediate and sometimes parochial citizen needs are emphasized and longer-term programs are ignored, deferred, or given short shrift. Local officials, particularly, have to be sensitive to social response to business or technical issues and must be sure to draw the balance between efficiency and human needs. As an example, in 1979 the elected officials in Hall County, Georgia, decided to contract for municipal fire services through a private firm and discontinue the municipal fire department. On the surface, the action seemed to be a step forward for the community: It provided a means to obtain fire services equal to or better than what they presently had, while controlling annual costs through a contracting agreement. Unfortunately, the commissioners misgauged the public response, and in a special recall election, all five members were deposed. The electorate had spoken: A wise business decision was a poor social decision. The commissioners' action was innovative; it just wasn't appropriate at that time and place.

Although the two-year election cycle produces responsive elected officials, it favors single issues and short-term gains and works against long-range decision making.

Organizational Structure

Local governments emphasize service delivery rather than long-range planning, and they emphasize immediate response rather than carefully calculated grand solutions. This emphasis on response produces administrators and personnel with an operational approach and places a premium on people who know how to get things done.

Most departments and agencies are long on doers and short on staff personnel, and large staffs that do not have day-to-day responsibilities are looked upon as frills in a departmental budget. They are particularly tough to justify to an official worried about getting reelected during a tight budget year.

The sheer range of local government activities also works against long-term decision making. With a wide variety of service delivery activities and most employees devoted to operational activity, there are few people who can take the long, broad view and fewer still who can invest the time required to pursue new approaches in their areas of responsibility.

The federalism of the 1970s brought with it, not only many new programs, but stringent requirements for citizen input and active citizen participation. As not-for-profit urban development authorities and health service agencies proliferated—ostensibly ensuring full participation in the provision of public goods and services by providers, receivers, and interested parties—the span of control from the general-purpose local government to the special entities got longer and funding became confused, diffused, and sometimes out of focus. Each board, commission, or other entity wanted to grow, in order to provide more services for its constituency for a variety of selfish and altruistic reasons. The amount of energy invested in communicating, coordinating, and keeping everyone informed became enormous, and the critical need was to keep all the organizations afloat with some degree of orderliness and cooperation. Under these circumstances, improving operations or instituting new approaches became a formidable challenge.

This management problem is compounded by the traditional diffusion of powers among local bodies that exists in most states. Overlapping responsibilities between cities and counties in provision of schools, police and fire protection, sanitary services, and other services is common. With the advent of tax reduction initiatives similar to California's Proposition 13, municipalities in several states are scrutinizing overlapping services and such issues as double taxation (city residents paying for services rendered simultaneously by the city and the county) much more carefully. Obtaining agreements to share services can be difficult when two competing service delivery organizations feel they are fighting for their lives. Often the competition leads to lack of cooperation and cumbersome artificial service delivery agreements that can decrease the efficiency and effectiveness of the services rendered.

The Fish Bowl Environment

The final impediment to effective long-term decision making is the fish bowl atmosphere that local government officials are required to operate in. Sunshine laws, which require full disclosure of the public's business, serve to discourage cronyism and corruption and permit taxpayers a better view of how local government activities are being conducted. While this complete public disclosure ensures that the locality is acting for the benefit of the taxpayer, it also provides an environment where mistakes can be discovered and magnified. Public officials operating in this environment have a very cautious attitude and tend to avoid activities that do not have a high probability of success. Trying new methods and equipment carries the risk of failure, and the electorate has traditionally had little tolerance for what it considers to be misspending of public money. The environment of full public disclosure and citizen concern about tax dollars thus discourage anything but a cautious no-risk approach.

The barriers to change in a local government can be substantial, but, as the next section points out, the public investment is also substantial and the cost of not implementing new technology can be high indeed.

LOCAL GOVERNMENT'S PLACE IN OUR ECONOMY

Government expenditures are an increasingly important part of the gross national product (GNP), the measure of the nation's total output of goods and services. While government expenditures at all levels are increasing faster than private sector expenditures, those at the state and local levels are growing even more rapidly than those at the federal level. From 1949 through 1976, federal expenditures rose from $41 billion to approximately $391 billion, an increase of 845 percent. At the same time, state and local expenditures rose from $18 billion to $117 billion, an increase of 927 percent. In terms of percentage of gross national product, federal expenditures rose from 16 to 23 percent, while state and local expenditures increased from 7 to 11 percent. In total, federal, state, and local expenditures accounted for 34.2 percent, or slightly more than one-third, of the GNP in 1976.*

Although state and local expenditures have grown more rapidly than federal expenditures since 1949, they appear to be dwarfed by total federal spending. However, examining the federal–state–local situation strictly on the basis of expenditures provides a distorted view.

Federal expenditures take place in several categories—among them, direct expenditures and transfer payments to other entities. The direct expenditures stimulate the economy directly, while the transfer payments are passthroughs to other entities that then spend the funds, making a contribution to the gross national product. These large passthroughs should more properly be counted as state and local expenditures when examining the relationship between federal funds and state and local funds.

Intergovernmental transfers—federal aid flowing to state and

* These figures were taken from Michael Bell and Richard Gabler, "Governmental Growth: An Intergovernmental Concern," *Intergovernmental Perspective,* Vol. 2, No. 4, Advisory Commission on Intergovernmental Relations, Fall 1976.

local governments—have increased dramatically since the mid-1950s. In 1954, federal aid amounted to 21.5 percent of state general revenues and 43.5 percent of local government revenues. By 1976, these percentages were estimated at 40.1 percent and 75.5 percent, respectively. So despite the fact that federal expenditures have grown rapidly, an increasing portion has been passed through to state and local governments.

Employment figures reflect this shift most dramatically. From 1949 through 1976, the number of persons employed by the federal government rose from 2.075 million to 2.85 million, an increase of 37.3 percent. During that same time period, the number of state and local government employees rose from 3.906 million to 12.227 million, a substantial 213 percent increase. State and local government, which accounted for 65 percent of the public sector employees in 1949, in 1976 accounted for 81 percent of the employment. Clearly, there has been an expansion in service delivery, employment, and expenditures by local governments since the late 1940s, and despite the difficulty in measuring its exact contribution of the gross national product, it can still be classified as substantial and influential.

Another measure of public sector growth is taxes. In 1953, the "average family" with an income of $5,000 paid 11.8 percent of its income in taxes. By 1975, the same family had an income of $14,000 and paid 22.7 percent of its income in taxes. There has been a substantial growth in government expenditures and taxes, and much of the growth has been at a local level. That growth has manifested itself in an increased range of all types of services that are technology susceptible. As local expenditures rise to a significant part of our total output of goods and services, it becomes absolutely essential to reap the advantages which technology can bring to the localities while at the same time anticipating and sidestepping undesired consequences.

THE NEED FOR ASSESSMENT

As has been discussed, the circumstances surrounding local governments work against long-term planning and decision making. Activities mandated by federal and state government often conflict with existing programs or policies and generally provide little room for accommodation. Internally, the election cycle, citizen demands, and the breadth of activity combine to maintain a focus on the here and now.

Over and above the internal and external pressures, there is a distrust of long-range planning. Some of the distrust results from vested interests fighting the exchange of their individual good for the common good, but much of the distrust results from a desire on the part of citizens to minimize government control and maintain a maximum of individual freedom. Planning in the longer range means, among other things, identifying those factors in the planning situation that are constant and those that are variable. By maximizing dependence on the constant factors and minimizing the variable factors, a more orderly process can be assured. If planning process, limiting its ability to respond to changing conditions and creating distrust and suspicion of the orderly consistency of results in the planning and decision-making process. The tendency to try to control is what produces rigidity in the planning process, limiting its ability to respond to changing conditions and creating distrust and suspicion of the orderly decision-making process.

On the other hand, local government is growing both in the type and span of services it delivers and in its impact on the economy in general. The limiting pressures we started to experience in the mid and late 1970s have produced a much tighter economic climate than the expansionist 1960s. Local governments are experiencing a combination of pressures: more demands for services, erosion of real income due to inflation, and the impact of slowly rising property-based taxes and significant citizen resistance to increasing taxes. The long period of increasing federal investment has just about run its course, as seen in such programs as revenue sharing; emergency assistance programs, such as the Economic Development Administration (EDA) countercyclical aid and the Comprehensive Employment and Training Act (CETA); and a host of grant programs. With federal revenues accounting for 75 percent or more of the total income for some communities, there is little prospect for substantial increase. In addition, with the federal government having budget difficulties of its own, the administration and the Congress have little interest in expanding existing programs or adding massive new subsidies.

Faced with these considerations, local governments will have a complex set of goals to attain. Not only will people have to be employed and services delivered, but both goals must be attained in circumstances that emphasize careful use of shrinking dollars and more service output per dollar spent. Although there are direct decisions that have to be made—increase taxes or decrease personnel or services, or somehow to muddle through—technology, properly employed, offers the greatest potential for increasing the efficiency and effectiveness of service delivery. As the cost and other consequences of employing technology rise, it becomes increasingly important to apply the assessment process to technology choices. There is also a need for flexible tools that will allow the planning and decision-making process to be carried out in a fashion that permits multiple choices and change of direction in overall thrust without changes in continuity. The assessment and forecasting process, properly carried out using some of the tools outlined in the next several chapters, will provide the decision-making process the flexibility that permits responsiveness and consistent, effective delivery of services.

CHAPTER 4

Forecasting and Assessment Tools

We have reviewed the precedents for, and the uses of, technology forecasting and assessment in government decision-making processes. We have also made the case for technology assessment as a most appropriate tool for local government decision making. Let us now turn attention to the tools that can be used to assist in the decision-making process. It was stated earlier that forecasting and assessment, although separable as disciplines, are intertwined. Thus, the tools that support one support the other. However, the tools are either primarily predictive or primarily descriptive in nature, with the predictive tools favoring forecasting and the descriptive tools favoring assessment.

Predictive tools are used to assist the forecasting process. Growth curves, which project the birth-through-death progress of a parameter that describes a technology, will be discussed in this chapter. Trend curves, which trace the progress of a family of technologies that each have individual growth curves, will also be discussed. Growth and trend curves suggest clear-cut empirical paths. For cases subject to more complex influences, the Delphi process, which seeks to maximize people's intuitive thinking processes and block out their biases, is another technique that will be reviewed.

Descriptive tools permit a better investigation of all aspects of a technology and thus serve the assessment process better than the forecasting process. Normative methods, discussed in Chapter 5, provide means to ensure that all elements in a process or system have been thoroughly analyzed. Analogies (Chapter 6) are probably the most powerful forecasting tool in that they consider factors other than technology in the evaluation of a given situation. They also present users with a pattern of thinking that is common to their everyday thought patterns—the evaluation of an unknown by comparison to one or more knowns. Analogies can be used to forecast or assess.

Normative methods, the Delphi process, and analogies can all be considered static methods of forecasting and assessment. Models, by contrast, are powerful dynamic forecasting and assessment tools. With the exception of growth curves, which are simplistic and single element oriented, the static methods look at present and future circumstances with very little emphasis on the path along the way. The model, however, simulates future performance or events through a given time span. By compressing time and expressing relationships with mathematical equations, it provides approximate responses to an evolving situation in a short time span with minimal resources. The model can be thought of as a continuous film of events as opposed to discrete snapshots of progress. Modeling is a powerful but immature tool, particularly for describing complex technological and sociological systems, and its advantages and shortcomings will be discussed in Chapter 7.

INFORMAL FORECASTING METHODS

It will be useful at this point to review the various types of informal forecasting methods, to shed more light on the usefulness of formal forecasting tools. The informal methods include the "window blind" approach, the "nothing is changing" approach, and the "expert opinion."

In the window blind approach, the forecaster assumes that everything will happen as quickly tomorrow as it happened yesterday and today. Progress in all areas will continue, and there will be no leveling off in the foreseeable future due to resource constraints or technology barriers. On the other hand, there will be no sudden advances in the pace of technology that will speed up the progress. A look at the United States' space program during its apogee in the 1960s and its decline in the 1970s provides proof that progress is not constant. In the 1960s, the United States commissioned a new generation of military aircraft, started development of a supersonic transport, and developed and perfected the systems and facilities that landed men on the moon. Within the industry, there was optimism and an expansive feeling. If the 1960s produced advances in technology, then the 1970s would produce more advancement and even more prosperity. The industry's unbridled optimism gave way to chaos when the Nixon administration canceled the supersonic transport and then significantly decreased space exploration activities by substantially reducing the budget of the National Aeronautics and Space Administration (NASA). The momentum that characterized the 1960s soon ebbed, and for the greater part of the last decade, the industry shrank to its previous size and level of attainment. The window blind had stopped. Clearly, nothing continues to grow at the same pace indefinitely without some external stimulus.

The second approach—nothing is changing—assumes that the status quo will prevail and that nothing will come along to disrupt the order of things. The freight transportation industry provides evidence that things do not remain the same. Long-haul cargo, once the exclusive province of rail lines and ships, is now transported through an elaborate combination of air freighters, piggyback vans that allow truck trailers to be carried by train or ship, and other mixed modes of carrier. There are plans to move coal by slurry pipeline, and the actual deregulation of air traffic along with the proposed deregulation of trucking will cause further changes. Any trucking company that forecasts no change to its business in this evolving environment will not survive the deregulation process. The rail industry, on the other hand, is looking forward to a revitalization. Fuel costs have made rail haul look more attractive, and the national shift from oil to coal will necessitate more rail transport capacity.

The window blind and nothing is changing forecasts may be viewed by some as no forecast at all, but they are forecasts nonetheless. One predicts a constant rate of progress; the other predicts no progress at all. There may be situations where those forecasts apply, but they will be few. A more rational forecast method for most situations would be an expert opinion forecast, where one person with significant knowledge of the past and present character of the technology at hand and an in-depth knowledge of all facets of the technology predicts the future course of events. This predictive capacity, built on experience and knowledge, is apt to define a succession of events that is a more accurate portrayal than a horizontal no-progress line or an inclined continuous-progress line. In many cases, an expert forecast is a valuable tool in the decision-making process. However, the expert opinion forecast is susceptible to the forecaster's biases, fears, hopes, and prejudices and will not meet the test of rigor that we expect from an analytical tool.

PREDICTIVE TOOLS

When substantial decisions about the future are to be made, it becomes desirable to invest time and effort in a structured process that provides a more accurate prediction of the course of events. Some of the tools that are useful in the predictive process, such as extrapolation, growth and trend curves, and the Delphi process, are described below.

Extrapolation

One way of understanding where we are going is to look back at where we have been; using information about our past rate of progress, we can project through the present and into the future. This process is called extrapolation. Figure 2 is an example of a simple straight-line extrapolation, where the line extended through the two points A and B produces a predicted point C in the future. This ultrasimple extrapolation assumes a convenient linear relationship between A, B, and C.

Figure 3 depicts a case that is more likely to occur, where the data points are scattered. Again in projecting into the future, a simple linear relationship between value and time is assumed. In this instance, the slope of the line is deterermined by calculating

FIGURE 2. Straight-line extrapolation.

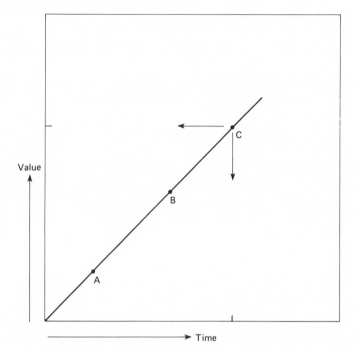

an average distance between the points and the line A–B, using least squares or a similar averaging technique.

The circled values are farther from the assumed trend line than the other values. The forecaster may choose to eliminate some of these data points because of other knowledge or because they are thought to be too inconsistent with other data to be relevant. If some of the data are removed, the line may shift lower or higher, but this treatment does little more than buttress a forecaster's intuitive opinion.

The Pearl Curve

Figures 2 and 3 represent ultrasimplified correlations, assuming relationships that seldom exist in practice. Figure 4 depicts a more sophisticated relationship taken from nature, the Pearl curve.

The curve is named after Raymond Pearl (1870–1950), an American biologist and demographer who made extensive studies

FIGURE 3. Extrapolation with scattered data.

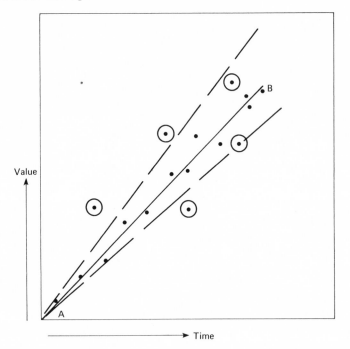

FIGURE 4. The Pearl curve.

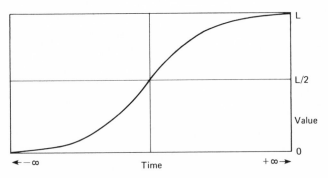

of the growth of organisms and populations. He found that cell and organism growth could be characterized by the equation:

$$y = \frac{L}{1 + ae^{-bt}}$$

where L represents the upper limit of growth of the variable y, t is the time variable, a and b are curve-fitting constants, and e is the natural logarithmic function. The resulting curve of y has a value of zero at minus infinity (the beginning of the time period in question), and y reaches the value L at plus infinity (the end of the time period in question). In actual use, some determination is made of the upper limit, L, by relating to such criteria as maximum growth, market saturation, or a similar limit. Then, on the basis of some historical data points, values of a and b are selected to provide a good curve fit. The finished curve can then be used as a predictor of future events.

There are precedents other than Pearl's work for assuming that the growth curve accurately depicts the natural growth of things. The Pearl curve describes how biological growth occurs in a closed environment, but there are physical and social environments where the laws of growth in a closed system apply as well. In an Otto cycle gasoline engine, for instance, the combustion reaction starts when a spark plug ignites the air–fuel mixture in its immediate vicinity. Relatively few molecules are involved in the initial reaction, and this activity is described by the left side of the growth curve, where growth occurs slowly. As the fuel–air mixture around the plug is ignited, it in turn ignites the adjacent molecules of air and fuel causing a flame front to form. Initially, the flame front is small, but as it grows it spreads rapidly, growing as depicted in the middle of the Pearl curve. As more and more molecules of air and fuel are ignited, fewer unconsumed molecules are left and the flame front slows down as depicted on the right side of the Pearl curve. As the reaction proceeds to completion, all molecules are involved in the combustion process and the upper limit L is reached.

A closed social system with limited membership can produce the same results, and a similar curve can be traced for the adoption of an innovation or a new approach. At first, during the ex-

perimental stage, few people will risk the adoption of a new process. If the few initial trials are successful, other people will hear
of it and the rate of adoption increases. As it increases, knowledge
of the innovation becomes more widespread and a large number of
adopters start to participate. This continues until a large percentage of the potential adopters have committed. At this time, the
growth curve starts to taper off until the theoretical limit is
reached, indicating that all potential adopters have committed to
the technology, and the adoption curve resembles the S shape of a
classic Pearl curve.

The most difficult part of the extrapolation technique is understanding where a particular technology is on the growth curve. If
the assumption is made that a particular technology is in the
ascent stage when in reality it is at the maximum growth point,
then a significant overestimation of the rate of growth has been
made and the forecast will be in error. On the other hand, if the
assumption is made that the growth of something has peaked
when it is still climbing, its growth has been underestimated.
Much of the art in growth curving is in understanding just where
a particular technology or situation lies on its own growth curve.

Trend Curves

We have just discussed the development of a growth curve that
focuses on a given technology; the growth curve, however, ignores
parallel and competing technologies. More often than not, a given
technology branches at some point during its birth-to-death cycle,
giving rise to a family of derivative technologies, each with its
own growth rate. If we are considering technology alternatives
rather than a narrowly defined technology, it would be more appropriate to look at a composite growth curve, a trend curve that
consists of the individual growth curves for all the members of the
technology family. The resultant trend curve takes its shape from
the maximum values of the individual growth curves.

For instance, when looking at the decrease in cost for computers, we see that the trend line for mainframes started to level off
in the 1970s, and it appears that the prospects for a significant
cost decrease will be minimal unless a dramatic breakthrough

can be made. However, if we take a look at the other newer members in the computer family, minicomputers and microprocessors, the future looks entirely different. The flattening trend line of the mainframe gives way to the lower cost trend of the minicomputer and the still lower cost of the microprocessor, extending the cost trend line even further downward.

To be sure, there are significant differences in processing power and flexibility as we descend downward along the cost (and technology) curve, but if the measure is computer cost over time, then the technology versus cost envelope obtained by extrapolating the cost trend through the three curves is far more representative than the single mainframe processor curve. And, in addition, technology is closing the gap, making the line of demarcation between a small mainframe and a large minicomputer almost indiscernible.

Growth and Trend Curve Parameters

The computer example also points up the need to be careful about the choice of measurement parameters. If the sole criterion is cost, then the cost curve is valid. If the criteria include cost and storage space capacity, or cost and processor speed, then the parameter chosen for the curve is clearly invalid. In order for the extrapolation process to be valid, the parameter we use as a measure of progress must have several characteristics: It must be an objective measure of a desired performance characteristic; it must describe the state of the art; it must apply to a wide range of technical devices that could be used to perform the function; and, finally, there must be enough historical data to permit some curves to be constructed.

Typical objective performance measures that we would see in municipal work might include the biological oxygen demand (BOD) removal efficiency of a wastewater treatment plant, the capacity of a refuse collection truck, or the memory capacity of a central computer. The common denominators for each of these parameters are that they relate to some desired performance, are measurable, and are applicable to all types and classes of equipment being measured.

The parameter must also reflect the state of the art. In the case of a refuse collection vehicle, a number of refuse volumes and compaction ratios may be available for a given size truck. Neither the volume nor the compaction ratio are true measures of the performance desired. The real measure of the state of the art is total weight of refuse carried for a given size truck. Similar conditions related to capacity and rate can be imposed for wastewater treatment plant efficiency, where increases in throughput affect efficiency, and for the computer, where dynamic core memory capacity is affected by the size of the operating or supervisory programs in use. In all these cases, the state-of-the-art measurement is really an efficiency term that combines capacity with throughput rate.

A parameter must also be general enough to apply to all the devices that can perform the desired function. Electric lamps serve as a good example, since they differ widely in configuration. For instance, lamps can use filaments, flourescent coatings, metallic vapor, or light-emitting diodes as their light energy source. The physical principles that govern their operation vary widely, but the ratio of light output energy (lumens) to electrical input energy (watts) provides a convenient measure of their performance, and light unit output in lumens per watt is a commonly used lighting efficiency term.

There must be a sufficient number of data points available over time to permit construction of a trend line that has some degree of confidence associated with it. Distortion of the data can result from measurement errors, which can occur when measuring the comparative performance of two competing or complementing technologies. Distortion can also be caused by external factors, such as a decision to provide accelerated funding to hasten the maturity of a technology (the Manhattan Project for atom bombs during World War II) or social pressures that inhibit technical development (the nuclear power backlash that started in the mid-1970s). The presence of many data points helps to develop a better picture of the average course of events so that anomalous data points can be recognized and culled out.

Finally, when various technologies are being compared, the data must depict them at the same level of maturity. Technologies can be in the developmental stage, in the early adoption stage, or at maximum growth or fully matured. It is important that competing technologies be compared at corresponding stages of development to ensure that compatible growth rates are being measured.

Limits of Extrapolation

Extrapolation uses past history as the basis for predicting future performance. It can be a relatively simple tool if the data are simple. Straight-line or modified straight-line projections are useful for limited time spans and stable rates of change. Growth curves assume growth rates based on relative maturity and are more difficult to construct, since some measurement of start and end points (technology life spans) and present position must be made in order to obtain a reasonable estimate of the rate of change that can be expected.

Trend curves are a series of growth curves of related technologies superimposed one upon the other as each of the technologies matures. They present an added complication of relating the levels of maturity of the various elements within the technology family.

Although trend and growth curves make use of the past to predict the future, they have a generalized simplistic view of life and are unable to factor in much else but historical progress. Extrapolation is particularly vulnerable to breakthroughs—those spikes in technology that produce great progress in an extremely short time, creating discontinuities in the simple straightforward curve of progress.

Extrapolation takes a deterministic view, examining only the exterior envelope of prior progress in order to project a future course of events. This approach considers only the gross progress of technology without looking at the internal events that shaped the progress. This simple projection approach can fall far short when assessment spans are short, or when assessments or forecasts become multifaceted issues requiring judgments about sociopolitical implications and their relationship to technology or about the trade-off of conflicting factors that interplay to aid or hinder growth. Such situations require a less rigorous, but more encompassing, approach.

EXPERT FORECASTING AND ASSESSMENT

The task of prediction and assessment in complex situations can be assumed by an expert forecaster—someone knowledgeable about the technology involved and the issues that shape it. An expert can make a qualitative evaluation of the circumstances as well as a quantitative evaluation of progress, providing a more encompassing, if more intuitive, forecast or assessment. However, an expert forecast or assessment is limited by the forecaster's biases, prejudices, fears, and desires as well as his span of knowledge. In short, it can be very subjective.

The shortcomings of individual forecasting or assessment can be overcome by gathering together a group of experts who have more than one point of view and who can ensure that each relevant issue is addressed. Ideally, the collective input of a group of experts can produce a broader, more balanced forecast or assessment. The problem with assembling a group of knowledgeable people is that what had been a purely technical process becomes a group social process and individual perspectives can be lost as group dynamics start to operate.

Good group results can be muddled by any one of several factors that crop up during a group meeting. One possibility is that a group discussion can be won by someone who expounds personal views forcefully and consistently, regardless of validity, because there is tremendous social pressure to submerge individual opinions in order to produce a group consensus. Since harmony is more important than opinion, a driving personality can overshadow good perceptions. Another problem with a group activity is that relative rank or authority can make a difference in the decision making. Although theoretically everyone's input should have equal weight, individual members of the group will sacrifice their opinions and be reluctant to contribute their knowledge if it conflicts with the opinions of an acknowledged expert or leader. Under these circumstances, authority or status commands a disproportionate share of the decision-making weight. Sometimes polarity sets in and members of the group freeze their positions so that no consensus can be arrived at. Sometimes members of the group clash over minor points, leading the discussion away from major issues. All these factors can and do combine to reduce the usefulness of group outputs. "Design by committee" is an all too familiar reference to the problems of group decision making.

The Delphi Process

The Delphi process, originally developed by researchers at the Rand Corporation, seeks to capitalize on the benefits that can be provided by an expert group, while minimizing the adverse side effects of the group process. Delphi uses two techniques—feedback and anonymity—to maximize the results of a forecast or assessment. Feedback is provided by summarizing group opinions of the issues in question and providing them to each individual panel member for review. The panel members read the group opinion, with supporting reasoning if necessary, and then revise or retain their own original estimates of the situation as the panel refines its judgment in successive iterations. Anonymity is provided by shielding individual opinions from the panel or group by summarizing group opinions and presenting individual reasoning without identifying the author of the reasoning. These two factors—feedback and anonymity—provide the benefits of group thinking without subjecting individuals to the social dynamics that usually accompany group activity.

The Procedure

In a Delphi process, a panel of experts is led through a series of questionnaires by a panel moderator. The panelists answer the questions, providing reasons for their answers. The answers are summarized with some of the reasoning included and provided to each member for review. The panel members again answer the questionnaires in a second round, taking into consideration the group's consensus. This answer–feedback–answer activity continues in successive rounds until a narrowing of opinion has been obtained or until there is little or no change in the group's opinion. At this point, the Delphi process has run its course, and the best possible group forecast or assessment has been obtained.

The phrasing of questions and the objectivity of the panel moderator are crucial to a successful Delphi. The questions must be framed so that they are clear and unambiguous, and capable of being answered without suppositions or qualifying remarks. They should be specific, rather than vague, and they should be objective so they do not lead toward a conclusion. The moderator must be careful to ensure that his role in the process is neutral. His role is to conduct the session, provide summarized results, and clear up ambiguities in the questions. He must concentrate on summarizing the results carefully and objectively, and he must ensure that his personal opinions do not bias the results.

Delphi results are strongly dependent on the quality of the panel members. Obviously, care must be taken to ensure that each member really is qualified to evaluate the technology in question. The selection process should also ensure that sufficient diversity is built into the panel to avoid a particular point of view or set of biases that would predetermine the outcome of the process.

A Delphi process can be conducted in a variety of ways. Members and the moderator can meet in a single room, or it can be conducted by telephone conference or by mail. Regardless of the communication medium used, the Delphi usually goes through several iterations or rounds.

The Delphi process begins with an orientation, followed by several rounds. During the orientation phase, the subject area and the goals of the Delphi are defined to all panel members. The methodology is explained, and questions about the administration of the Delphi are answered.

In the first round, the questions are fairly general and not too tightly structured. The panelists are asked to make an assessment or forecast about the area in question. The panelists are experts in their area and should be more knowledgeable than the moderator. Consequently, during the open structure of the first round, issues may surface that have not been previously considered but that may have significant bearing on the subject matter. The panelists' information is submitted to the moderator, who then consolidates the information and develops a coherent set of questions for the second round.

In the second round, the moderator presents the consolidated results of the first round to the panel members. The panelists are asked to provide new answers in light of the group information received, as well as reasons for their answers. The answers are again collated by the moderator, who provides a statistical summary of the results, listing the arguments for and against the majority opinion.

In the third round, the panelists are again provided with the group statistical information and opinions that support or contradict the majority opinion. Those views that differ significantly from the majority are listed, providing arguments against the majority opinion just as would be done in a face-to-face discussion. However, anonymity is preserved, and confrontations are avoided.

In the fourth round, the process is repeated again. At this point in the Delphi, convergence is usually reached, or there may be clear indications that it will not be reached, and the process can be terminated. If, on the other hand, the moderator sees a continuing shift in opinions, he may ask for more justifications for positions taken and carry the process through one or more additional steps before the Delphi is concluded. At this time, those answers that appear consistently from round to round should be construed as conclusions. After the final round, results are summarized and presented as the aggregate opinion of the panel.

A Demonstration

As a demonstration exercise, I conducted a truncated Delphi process for two graduate-level classes in technology forecasting and assessment over a two-year period in 1977 and 1978. Solar energy was chosen as a topic, since it had received much publicity in the 1970s and is a clearly identifiable technology, easy to understand.

The Delphi process was truncated in order to save instruction time. The questions were structured at the outset, and the process was terminated after two rounds. However, despite the limitations—the background of the "experts," the highly structured questions, and the limited number of rounds—some consensus evolved, even among two Delphi sessions conducted in different years. Here is the Delphi questionnaire that was used.

Delphi Questionnaire—Solar Energy

Solar energy is a much talked about solution to the present energy crunch in the United States. The purpose of this questionnaire is to elicit your opinion, as one member of an expert Delphi panel on solar energy, as to its prospects for future use in our society.

Please answer the following questions and provide brief justification for your position.

1. When will solar energy be widely used (50% for new construction, 5–10% for retrofit) for heating purposes?

    _____ 1980
    _____ 1990
    _____ 2000
    _____ Never

    Why?

2. Who will be the principal user (implementor) of solar heating systems?

    _____ Individuals
    _____ Residential (home) building contractors
    _____ Industrial contractors (manufacturers)
    _____ Government

    Why?

3. What will be the principal use for solar heating?

    _____ Hot water heating
    _____ Winter heating
    _____ Summer air conditioning
    _____ Industrial process (for example, paint drying, tobacco curing)

    Why?

4. When will solar to electrical conversion systems be widely
   used? (30% for new uses, 20% or more for existing uses)

   _____ 1980
   _____ 1990
   _____ 2000
   _____ Never

   Why?

5. Who will be the principal users of these systems?

   _____ Individuals
   _____ Commercial users
   _____ Utilities
   _____ Government

   Why?

6. What will the principal use be?

   _____ Central utility power generation
   _____ Spot utility power augmentation
   _____ Direct conversion by users

   Why?

7. The principal driving force for the use of solar energy over the
   next 30 years will be:

   _____ Shortage of fossil fuels
   _____ Cost of fossil fuels (i.e., cartel action by OPEC)
   _____ Federal policy
   _____ Technology advances

   Why?

The questions with the strongest and most consistent responses
were the ones relating to when solar technology would be in widespread use. These specific forecast questions were the easiest to
answer.

Question 7, which asked what the driving force for the adoption
of solar technology would be, was an assessment question. The

TABLE 1. Delphi responses, %. (Majority opinion in boldface type.)

Question		First Delphi		Second Delphi	
		Round 1	Round 2	Round 1	Round 2
1. Solar heating in wide use.					
1980	A	8	8	0	0
1990	B	31	16	**33**	20
2000	C	**61**	**76**	**33**	**80**
Never	D	0	0	**33**	0
2. Principal user of solar heat.					
Individuals	A	**30**	**34**	**33**	20
Building contractors	B	**30**	22	**33**	**60**
Manufacturers	C	23	28	17	0
Government	D	17	16	17	20
3. Principal use for solar heat.					
Hot water	A	22	21	30	33
Winter heat	B	**44**	**53**	**50**	**67**
Air conditioning	C	17	16	10	0
Industrial processes	D	17	10	10	0
4. Solar electric in wide use.					
1980	A	0	0	0	0
1990	B	15	18	33	0
2000	C	**77**	**73**	**50**	**80**
Never	D	8	9	17	20
5. Principal users of solar electric.					
Individuals	A	25	21	0	**50**
Commercial users	B	25	29	33	0
Utilities	C	**37**	**50**	**50**	**50**
Government	D	13	0	17	0

TABLE 1. (*Continued.*)

Question	First Delphi		Second Delphi	
	Round 1	Round 2	Round 1	Round 2
6. Principal uses of solar electric.				
Central utility power A	27	29	33	20
Spot utility use B	**53**	**50**	**50**	**60**
Direct conversion C	20	21	17	20
7. Principal driving force for use of solar energy.				
Fossil fuel shortage A	**65**	**65**	17	20
Cost of fossil fuels B	7	7	**50**	**40**
Federal policy C	14	28	0	20
Technology advances D	14	0	33	20

chosen reason differed in both sessions, and further rounds would have been required to produce more convergence. Interestingly, solar electrical energy, which is less mature than solar thermal energy, fared just as well in consistency of results. The two Delphi exercises produced interesting and useful convergence in their answers (see Table 1), demonstrating the value of the Delphi process for predicting the path of progress in complex social–technical issues.

CHAPTER 5

Normative Methods

Trend curves and Delphi are predictive in nature. Although there is descriptive information standing behind the construction of a trend curve, the curve itself conveys little about the technology it represents, other than information about the chosen growth parameter. The Delphi process, by its nature, is more descriptive, but Delphi usually focuses on many of the implicit factors associated with a technology rather than on the shape or structure of the technology.

In some instances, particularly when an assessment is being made, descriptive information that provides some idea as to the dimensions or intervals of a technology can be of more value than predictive information. Normative methods perform the descriptive function that enables us to understand more completely the technology at hand, allowing us to draw some better conclusions about the consequences of its use or to make projections about its evolution.

Normative methods map out the elements of a system and examine the relationships between those elements. They can be used to assure a complete description of a tool, process, or organization and to establish goals for the elements involved.

The three most common normative tools are relevance trees, morphological models, and flow diagrams. Relevance trees are used when hierarchy is involved. They thus form an "organization chart" for the elements in a process or system. Morphological models are used where a system or process can be broken into independent elements, all of which function at the same level. They are valuable for describing and analyzing multiple elements within a system or process that must interrelate with one another. Flow diagrams are used when the elements of a system or process have a time-sensitive relationship and can be described in terms of paths or sequential steps.

RELEVANCE TREES

A relevance tree is used to break down a process from its central component or function into the successively finer components that support it. Figure 5 shows a simple relevance tree with three levels. Level A, the topmost function, is broken into supporting elements B1 and B2, which are further divided into supporting elements C11, C12 and C21, C22, respectively. The point at which each element branches into subelements is defined as a *node*. A node always has two or more supporting subelements, but there is no maximum and no requirement that each node have the same number of supporting elements. The criteria for the elements are that they identify an independent function not overlapping with other elements and that the sum of all elements tied to a node describe all the possibilities or characteristics of that node.

FIGURE 5. Simple relevance tree with three levels.

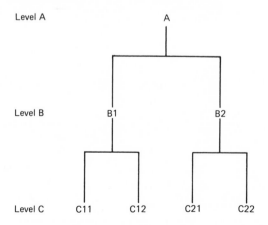

Figure 6 is a relevance tree delineating the primary functions a building inspection organization might perform. The end goal of the department is to ensure that all industrial, commercial, and residential structures are built and maintained in a manner that will provide safety for the citizens who live in them, work in them, or visit them.

In order to accomplish that goal, localities generally exercise control over all aspects of the building design and construction

FIGURE 6. Relevance tree for the primary functions of a building inspection department.

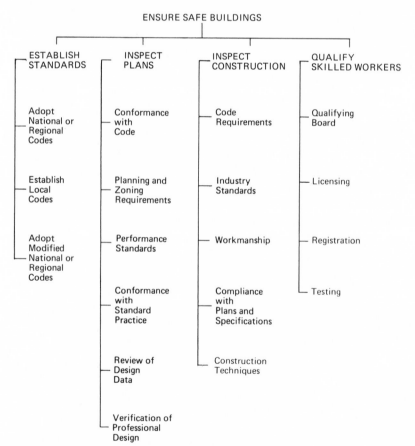

process. The control starts with the establishment of standards governing how a building is to be designed and built. The standards are reflected in plans and specifications that describe the configuration of a building and its equipment in substantial detail, incorporating both the requirements of the owner and the community at large. The plans and specifications are used as construction documents and as historical records of the building configuration.

Design is followed by construction. During this phase, skilled workers interpret the plans and specifications, using judgment and knowledge of standard construction practices to carry out the intent of the locality, the owner, and the designer. Buildings are not mass-constructed items, and few buildings are identical in design and equipment. As a result, skilled workers can and do use considerable latitude during the construction process.

To meet the end goal, it is necessary to control each of the subelements in the design and construction process. These subelements under the general goal are the establishment of standards, the inspection of plans, the inspection of construction, and the qualification of skilled workers. All the subelements are necessary to ensure safe structures, but choice and emphasis may vary from subelement to subelement, and indeed within the subelements themselves.

In the area of standards, a community may choose either to adopt widely accepted regional or national codes or to establish its own codes for structures, plumbing, electrical power, heating ventilation and air conditioning, and fire protection. In some cases, the state may mandate the codes that must be used. If a locality has a special set of needs that are not covered by existing codes, it may opt for a combination of national or regional codes modified for local needs. Typical areas for specialized codes may be the rehabilitation of old housing stock, specialized fire safety requirements for high-density zoning, or solar energy provisions. The question of code adoption is governed by state mandates, historical attitudes, community size, and skills levels within the local government itself.

Plan inspection, an activity that derives from code adoption, provides for a review of building plans and specifications before construction to ensure that code requirements have been properly translated into building design. The plan reviewer can evaluate in detail the design calculations and drawings to ensure that traditional code requirements, and perhaps new energy and occupancy performance standards, have been met. This is often done in localities that have sophisticated code requirements or dense high-rise structures. On the other hand, the reviewer can compare drawings and specifications with accepted standards or simply conduct a cursory inspection that establishes that the design was done by a knowledgeable professional architect or engineer. The last approach is most practical in a small community where the contact between designer and reviewer is more personal.

The plan review can also incorporate a review for compliance with zoning ordinances (setbacks, utilities, streets) and any other special conditions imposed during the planning process.

The actual construction inspection emphasizes code compliance, workmanship, compliance with the plans, construction practices, the use of industry standard materials or equipment, or any combination of activities depending on the structure of the inspection department, the training and aptitude of the inspectors, and the general approach to inspections.

As with the plan review process, construction inspection procedures vary widely. They range from detailed on-site activity, where a resident inspector reviews test data, witnesses major construction and installation tasks, reviews equipment for conformance with standards and codes, and approves field changes, to visual inspections at key points during construction or a walk-through inspection to certify fitness for occupancy.

The licensing process, which is established to ensure that work is performed by competent people, can be accomplished by any one of a number of procedures: Communities can use state certifications; they can establish an examining board staffed by experienced skilled workers; they can develop formal written and practical qualification tests; they can accept certification by local craft unions; they can issue business licenses as certification vehicles.

Not all the means used to ensure safe building inspection need the same degree of emphasis. One community may decide to place great emphasis on code requirements and rely heavily on the skills of professional designers to carry out the code requirements. Another community may emphasize on-site inspection and the use of standard construction industry practices. Still another may place the responsibility with skilled licensed workers, with less emphasis on the code, design, and inspection aspects.

WEIGHTED RELEVANCE TREES

The relative importance of each task determines where emphasis should be placed. A weighted relevance tree provides an excellent vehicle for organizing and visualizing a priority-setting process. A weighted relevance tree can be developed from an existing tree by adding numerical values that reflect the relative importance of each element or subelement. For ease of use, the numerical values of each group of elements or subelements are normalized so that they add up to one. By using a simple multiplication process, the relative importance of any element or subelement can be calculated.

To illustrate the process, let's assume that a local government's inspection department is going to be reorganized to meet changing community requirements. The tree developed in Figure 6 makes a convenient starting point. It lists all the relevant functions the department has to perform, and it can be used as a tool during a review with concerned officials to determine where emphasis should be placed. The review would consider citizen attitudes, changing inspection priorities brought on perhaps by new energy conservation standards, available skills, and budget constraints in order to establish departmental objectives. The objectives would then be translated into relative emphasis to be placed on each element and subelement.

Figure 7 shows one possible outcome. The major functions of construction inspection and plan review received greater priority than establishing standards and qualifying skilled workers, according to the numerical values assigned to them. Note that the numerical values of the four tasks at the first level are normalized so that they add up to one. Construction inspection is targeted to receive 40 percent of the departmental emphasis, plan inspection

FIGURE 7. Weighted relevance tree showing priorities of a building inspection department.

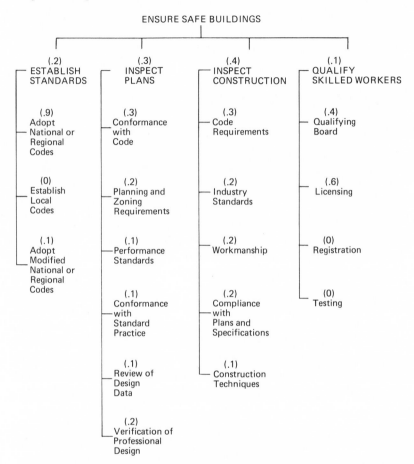

receives 30 percent, establishing standards receives 20 percent, and qualifying skilled workers receives 10 percent. The major tasks can now be further divided into subtasks. For instance, if it were desirable to place the highest priority on the adoption of a regional code and making the workers' certification requirements more flexible, then those two elements would receive more emphasis than that shown in Figure 7.

If all the subelements within an element are assigned normalized values (the total equaling one), then the total value of the subelement to the department can be analyzed by multiplying the element values by the subelement values. For example, it has been determined that conformance with codes should receive 30 percent of the emphasis within the plan inspection task. Plan inspection in turn should rate about 30 percent of the total departmental effort. Inspection of plans and specifications for conformance with codes should then receive .3 × .3 = .09, or about 10 percent. This percentage can be used as a guide in determining how much of the overall departmental resources (budget and personnel) should be invested in meeting this requirement.

As the preceding example points out, relevance trees can be used to point out hierarchical relationships and to ensure a complete description of functions. In addition, when the elements and subelements are assigned normalized numerical values, the tree can be a handy tool for evaluating priorities and allocating resources.

MORPHOLOGICAL MODELS

The relevance tree discussed in the previous pages is useful for describing hierarchical relationships where some elements are subordinate to others. There are many instances, however, where all the elements are parallel and contribute equally. Morphological models are useful for characterizing these horizontal relationships. In municipal operations, vehicles and process configurations can easily be examined and described by the use of a morphological model. In this section, we will use configurations for refuse collection trucks and alternatives for wastewater treatment processes to illustrate the usefulness of morphological models.

Figure 8 depicts the major elements that would be considered in the selection of a municipal refuse collection truck. Three elements—engine, transmission, and rear axle—are part of the drive train. Three elements—loader configuration, compaction method, and compactor drive—are part of the collection and storage function. Of the two remaining elements, cab configuration

FIGURE 8. Morphological model for selecting a refuse collection truck.

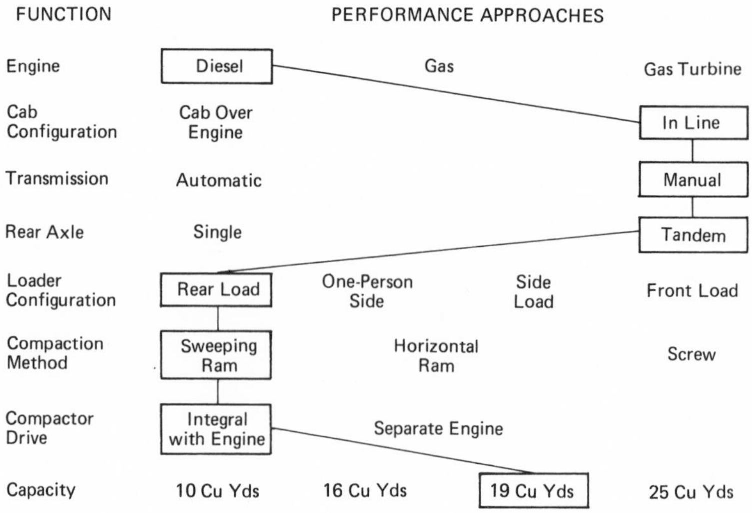

affects overall length, and thus maneuverability, while capacity affects overall vehicle size and gross weight. There are a total of 576 conceivable combinations contained in Figure 8. Each option in theory is separable and not dependent on the other, and therefore a morphological or shape-oriented framework is best suited for this kind of analysis. In reality, some options are dictated by local collection practices, street configurations, and ordinances, and the selection of one option strongly influences the selection of another. Two typical situations, collection in a densely populated urban area and collection in a more sparsely populated suburban area, will show these relationships.

A refuse collection route in a densely populated area with a nearby landfill could place a premium on a small, maneuverable vehicle. This would lead to the selection of a single rear axle and cab-over-engine configuration with a small payload capacity. The small carrying capacity and resulting efficiency decrease could be offset by using a one-person side-load configuration if street width permits side loading. Driver–loader work load can be reduced by using an automatic transmission, and a gas engine might be chosen over a diesel because of smaller size and weight and greater flexibility. The gas turbine could conceivably fit well in this configuration because of its light weight and high power output, but turbines have high fuel consumption during idle and part-load operations. These characteristics make turbines better suited for long-haul, over-the-road operations, where they are now finding their first use.

An entirely different situation, a suburban route served by a distant landfill, might be better answered by the solid-line configuration illustrated in Figure 8. In this instance, the landfill travel distance emphasizes payload, and the usual suburban street layout permits a larger, longer vehicle. For this situation, a large-capacity compactor would be chosen and a tandem rear axle would be used to keep the vehicle within the weight limitations specified by the state or locality. The large size and weight dictate the use of a diesel engine, since large gasoline engines are unable to meet present-day exhaust gas emission standards and are now all but nonexistent. The combination of high capacity and more distance between stops would make a multiperson-crew with rear loading a good option. If street width and parking regulations permitted, side loading could be chosen. A sweeping ram, which provides better compaction, would be chosen to maximize payload in this application.

Since size is not critical, an in-line cab and engine configuration could be chosen because of its greater simplicity and ease of maintenance. The driver, with less activity than in a single-person operation, could easily tolerate the additional work load of a manual transmission in exchange for its greater flexibility. With the driver monitoring the equipment operation, the compactor could also be driven by the engine used to propel the truck in this application. The gas turbine could also be matched to this application, although its torque characteristics require an automatic transmission.

Decisions about diesel or gasoline engines, transmission type, compactor type, and compactor drive will be influenced by such factors as manufacturer's options, cost, reliability, maintenance skills, and local preferences. Even though in many cases features are interrelated, the construction of a morphological model allows a deliberate step-by-step process to replace the intuitive selection process. This permits a trade-off between items that are influenced by local conditions and those that provide lower cost or higher efficiency. It also permits consideration of some combinations that may at first appear to be unorthodox but that hold real potential for system improvement.

Morphological models can be used to evaluate processes as well as equipment. Figure 9 delineates some of the options available to a wastewater treatment plant designer. The fundamental objective is to remove the wastes from the water stream, concentrate them, and dispose of them in a nonharmful manner. The water that carries the waste itself must be clean and free of contaminants that could be harmful to life when the water is returned to the environment.

The basic processes used to treat the wastewater are pretreatment, waste decomposition, and water disinfection. Waste products from the first two processes, a mixture of water and solids called *sludge,* are further treated before they are disposed of.

The initial processes—screening, grit removal, and primary sedimentation—remove natural and man-made materials that will interfere with the waste material decomposition process. Screening and grit removal remove such natural materials as dirt

FIGURE 9. Morphological model for evaluating wastewater treatment processes.

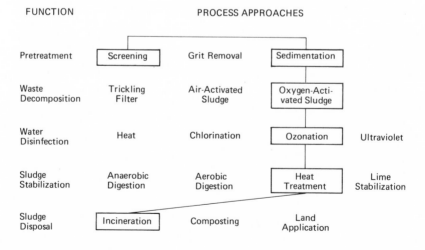

and rocks, and such man-made materials as metal, plastic, and cloth. Sedimentation removes large particles of material that are easy to handle and can be routed to a disposal process. These operations are primarily mechanical removal processes and vary only in the details of design and construction that make them best suited for the treatment site and waste to be handled.

Biological processes are used to convert the complex organic materials present in the wastewater stream into simpler, more benign materials. They use natural microorganisms that decompose the materials by using them for food in their own growth processes. The microorganisms break down the complex organic materials into basic gases and cell material. The gases evolve out, and the microorganisms eventually die, producing the sludge by-product that must be further treated. Nonbiological (physical–chemical) processes are also used to break down organic material, but they are less widely used than biological treatment processes.

Three commonly used biological processes are depicted in Figure 9. The trickling filter uses stones, wood, or plastic to provide a surface area that organisms can cling to and grow on. As the wastewater passes through the media, the microorganisms form in a slime and feed on the wastes in the water stream that serve as nutrients. The slime then sloughs off the media, forming sludge, which is then removed, and a new layer forms to continue the process.

The air-activated sludge process is somewhat the same, except that the organisms are suspended in the wastewater, and their growth is supported by passing oxygen in the form of air through the water–microorganism mixture. The sludge, which contains live and dead organisms, drops to the bottom of the tank that holds the water being treated. Some of the sludge is withdrawn for disposal. The remainder is recirculated to stimulate the biological growth process.

A derivative of the air-activated sludge process, the oxygen-activated sludge process, differs in that pure oxygen is substituted for air, increasing the oxygen content of the aeration gas by a factor of five and increasing biological activity.

The trickling filter process is simple in function and has found widespread use in communities where a less stringent treatment process can be tolerated. The air-activated sludge process is more complex but removes contaminants more efficiently and takes up less space than a trickling filter. The oxygen aeration process is best suited for areas that have strong wastes, high waste variation, or limited treatment plant space.

The next step in the process is disinfection, the destruction of disease-carrying organisms present in wastewater. The most common method of disinfection is chemical treatment by chlorination. Chlorine is a strong oxidant that chemically alters the life processes of disease-causing organisms. However, there is much concern over the possibility that chlorine produces carcinogenic compounds, and the search is on for other ways to disinfect wastewater.

One alternative is ozone, a highly reactive form of oxygen usually formed by producing a high-voltage discharge in oxygen. Ozone is a strong oxidant that functions the same way chlorine does, but it rapidly decomposes back to oxygen when it is added to water and thus produces no chemical residual. When examining the combined biodegradation and disinfection processes, it is evident that the same plant that produces oxygen for the oxygen-activated sludge process can also be used to provide ozone, thus presenting a very cost-effective process match.

Heating water to its boiling point will destroy major disease-producing bacteria, but the energy requirements for such a process make it impractical for a wastewater treatment. Ultraviolet energy, either from the sun or from an artificial ultraviolet source, is also a good disinfectant. However, the suspended matter in the wastewater and the water itself will absorb ultraviolet radiation. Therefore, extreme care must be taken to ensure that the ultraviolet penetrates the wastewater stream under all circumstances. This consideration severely limits its utility for a wastewater treatment plant.

The sludge produced by the pretreatment and biological decomposition processes contains concentrations of the materials that make the wastewater offensive; it also contains the pathogens that threaten our health. For these reasons, sludge must be further treated before it is ultimately disposed of. Figure 9 shows four major potential approaches: anaerobic digestion, aerobic digestion, heat treatment, and lime stabilization.

The two methods of digestion are an extension of the biological processes in that the sludge is further decomposed at elevated temperature over an extended time period by microorganisms. If the biological process takes place without oxygen, then the digestion is anaerobic and a by-product fuel, methane gas, is produced. If the process uses air, then it is aerobic and carbon dioxide is produced as a gas.

In heat treatment, sludge is heated in a pressure vessel to destroy pathogens and accelerate thermal decomposition. This treatment also decomposes cell walls, permitting efficient separation of water and solids.

Lime stabilization raises the pH of the sludge, a situation that will prevent microorganism growth. As a result the sludge will not create odors or health hazards. However, the sludge must be disposed of before the pH drops, since the organic matter that permits microorganism growth is still present in the sludge.

The final treatment processes are composting, where the sludge undergoes further biological destruction to a stable end product; thermal reduction (incineration), a process that reduces the sludge to an inert ash; and land application, where the sludge is spread on soil to take advantage of the nutrients it contains. Again, as with the other processes, there are natural matches.

Composting can be effectively used where the sludge has had limited stabilization and can still undergo biological decomposition. Incineration requires extensive dewatering to reduce energy requirements and benefits from the high organic content that can be provided by heat treatment. Incineration is a hardware-intensive industrial process usually employed in a sophisticated plant that a large city must build because of limited land for facilities and disposal. It thus makes a good match with either of the activated sludge aeration processes.

Land application carries with it a potential for pathogen carry-over, and its use is presently limited to fertilizing fields for forage crops, which will not show up directly in the food chain. For these reasons, digestion would make an effective stabilization process for land application. Although the processes listed in Figure 9 can be matched in many combinations, including chains of processes, the configuration shown delineates a possible high-technology alternative that is process intensive and well suited for a large city.

FLOW DIAGRAMS

Relevance trees and morphological models are well adapted for analyzing systems and processes, but not for describing processes that have a time dimension or proceed in steps or sequence. The flow diagram is better suited for such situations.

The flow diagram maps all the routes or paths by which a task can be accomplished. This allows the identification of difficulties or costs associated with each alternate path and makes it possible to identify new alternatives that can be evaluated on the same basis as the original paths.

Figure 10 shows the possible paths that could be used to transmit an emergency call for assistance to a public safety communications center. The most usual transmission path is through the telephone, using an emergency number. The number 911 has been endorsed nationally as an emergency response number for universal use by public safety agencies. Universal adoption of this three-digit distinctive number would minimize time delays in calling for emergency services. The use of the number, however, requires telephone operating companies to install special equipment to route 911 numbers to emergency service agencies in areas with more than one central telephone office, and the telephone operating companies have not moved rapidly to initiate 911. In addition, to be effective, a 911 number must place police, fire, and emergency medical services at the caller's disposal through a central switching point, and this requires the concurrence of all the affected local agencies. In most cases, it involves multiple police, fire, and emergency service departments, and each jurisdiction must give up autonomy in order to provide efficient response. Such interlocal agreements are difficult to develop, and 911 has not moved rapidly into service as a result.

In areas where 911 has not been instituted, jurisdictions assign one or more seven-digit telephone numbers for emergency response. These numbers vary from jurisdiction to jurisdiction and can be difficult to remember in times of stress, particularly when separate numbers are assigned for police, fire, and emergency services. Despite their shortcomings, the seven-digit numbers are widely used and provide the most frequent interface between citizen and emergency services.

FIGURE 10. Flow diagram for emergency call response.

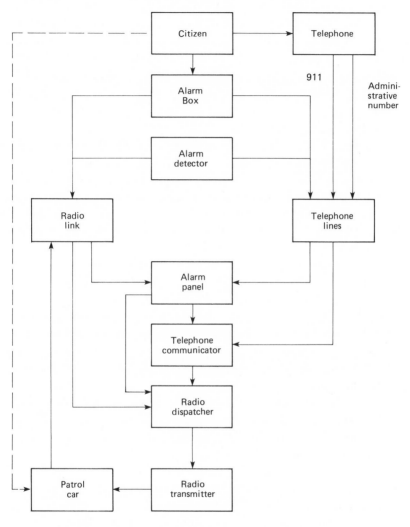

In the situation depicted in Figure 10, the citizen's call goes to a communicator, who assigns it to a dispatcher for radio or alarm dispatch to an emergency service vehicle. Automatic fire or burglar alarms follow a different path. The alarms are often tied to a communications center by a telephone line, but in some circumstances, this function can be better accomplished by using a radio transmitter at the alarm site and a corresponding radio receiver in the communications center. The alarm panel can be located near the communicator, the radio dispatcher, or both, and the alarm itself can be a flashing light, a tone designating a problem at a predetermined location, or even a recorded voice alarm. The alarm transmits its signal rapidly and can provide warning without human intervention. However, this type of alarm provides no supplemental information and can give misinformation if a malfunction causes a false trip.

The final link, patrol car to dispatcher, is a happenstance link. It depends on the presence of a patrol car in the immediate vicinity of the incident or on some previous notification of a problem. Although it is an efficient link—a trained professional is communicating pertinent information over a dedicated communications channel—it cannot be considered a primary source of emergency calls.

An analysis of the flow diagram provides some alternate paths. If cost is a consideration in alarm systems, telephone dialers linked to each of several alarms can be attached to a single telephone line. When an alarm is activated, the dialer automatically dials a predesignated emergency number, saving the cost of a dedicated telephone line. On the other hand, if maximum reliability and security are desired, each alarm may be provided with a dedicated full-time telephone line. Some cities use emergency call boxes with dedicated communications lines for emergency calls. Although this approach is more expensive than using the commercial telephone system, it provides a secure response and tells the emergency service where the call is coming from.

Some telephone companies are now providing extended 911 service, which automatically provides a dispatcher with the name and address of the person or organization listed in telephone company records under the calling telephone number. This shortens the response process significantly by decreasing the amount of information an emergency caller must provide.

The radio link shown consists of a high-powered transmitter-receiver (the base station) transmitting to a low-powered mobile or portable radio. This situation provides good outbound (base station to mobile or portable radio) transmission but unsure inbound capacity. In order to increase the reliability of inbound transmissions, satellite receivers can be located in various areas in a jurisdiction and connected back to the base station by telephone cable. Spotting the receivers in accordance with a predetermined coverage pattern assures that the low-powered mobile or portable radios will always have a receiver in sight, thereby increasing the reliability of inbound message transmissions.

An examination of the communicator–dispatcher link suggests other alternatives. Instead of dividing tasks between telephone answering and radio transmission, a locality can be divided into geographic sectors with a dedicated telephone communicator-radio dispatcher assigned to each sector.

Community service centers are a desirable means of providing citizen services. They put service workers directly in touch with citizens in a neighborhood setting, and the smaller scale of local service centers avoids the institutional feel of a centralized service delivery center. When a neighborhood center is located, accessibility—and therefore transportation—is an important criterion. Figure 11 shows a flow diagram that can be used to analyze the transportation modes to such a center.

The paths on the left side of the diagram are personal means of transportation; the paths on the right side are public transportation. The most limited form of transportation is walking. In order to be within walking distance, a service center would have to be located a mile or two from the majority of citizens served. If the sick or elderly are receiving services at the center, walking would not be possible and other means are necessary. The next alternative, bicycling, extends the range to a few miles but is limited to those citizens who are young and in good health.

The automobile is the primary means of transportation in suburban areas, where mass transit systems are not extensive. If a service center is to support citizens who drive, parking facilities would be required. A service center that can be reached only by car does not provide accessibility to people who cannot afford or who are unable to drive. Since such people would probably be a substantial percentage of the people receiving services, the centers should be located near public transportation.

Buses are a very flexible form of transit. They are inexpensive

FIGURE 11. Flow diagram for transportation to a community service center.

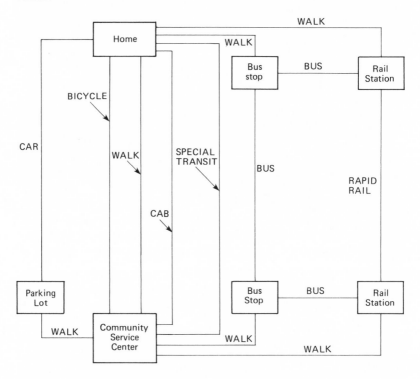

compared to other forms of vehicular transit and stop often enough along their routes to provide a short walking distance from a home to a bus stop and from a bus stop to a service center. Since bus service is a public or a quasi-public service, bus routes can usually be tailored with little effort to provide maximum access to a community center.

Rapid rail transit is faster than bus transit over long distances but has fewer, rigidly fixed stop points. Often a rapid rail must be supplemented by a bus connection at either or both ends of the trip. Rapid rail provides maximum accessibility particularly in an urban area if the service center can be located next to a station.

For people who do not have access to a car or mass transit, the most expensive form of transportation, door-to-door service by taxi cab, must be used. This premium transportation falls in the "last resort" category because of cost.

In locating a service center, there are several choices. Where possible, the service center should be located at the centroid of the population group it serves, in an area conveniently accessible to mass transit. If a good transit system exists, and the center serves a decentralized population, then the emphasis can be placed on location close to a transit hub. When these criteria cannot be met, or where a disabled or handicapped population is being served, many localities have substituted a special transit service, subsidized and run by the community. This service generally uses vans, smaller than buses but larger than cabs, to provide transit on a scheduled, dial-a-ride, or other on-demand basis.

The flow diagram can be used effectively in any area of transportation or communications where a comprehensive evaluation of all possible paths is desired. It can be used to evaluate an existing system or to ensure completeness in the design of a new system.

Normative methods help organize and structure a problem so that all alternatives may be weighed, but they can add too much rigidity to the solution process if not carefully handled. In addition, in the case of the relevance tree, the numbers have a tendency to create their own validity. However, when used judiciously as an organized element in a comprehensive assessment of a given situation, normative methods can be of great value.

CHAPTER 6

Analogies

As stated earlier, the process of technology assessment carries with it some aspects of prediction and some aspects of measurement. The predictive side of assessment looks at where a given technology or group of technologies will be in the future. After all, that is what we are planning for and anticipating. The predicted path is then used for the measurement process. Given that we will have this technology in place, what will be the impact on people using it? Who will benefit from it? What will it cost the implementing organization? What will it do for society in general? Each of these questions is meant to assess the impact of a technology, to measure the cost and benefits of its implementation.

Growth curves and trend curves focus on extending previous history forward on an empirical basis, using growth itself as a primary measuring index. Delphi is an attempt to broaden the focus by consolidating expert opinions, but the major emphasis is on the technology, and consideration of external factors comes as a by-product to the basic process. The normative methods described in Chapter 5 help characterize and identify all the elements of a technology or process to ensure completeness of the evaluation process. However, these methods add little to our knowledge of where we are going or what the effects of a new technology will be. In some respects, the analogy is a more comprehensive assessment technique than any of the others previously mentioned, since it attempts to look beyond straightforward technical issues and places technology in context with our society. It certainly is more consistent with the fluid give-and-take of a local government situation in that it organizes and categorizes our perceptions in the areas of technology, economics, and society, in order to measure the utility, and thus the potential for success, of a given technology.

The process of analogy is described in the Oxford English Dictionary as "presumptive reasoning based on the assumption that if some things have similar attributes, their other attributes will be similar." Martino, quoting O'Connor, describes a formal analogy as a "serious attempt to uncover other similarities (between two events) once some basic resemblances are noted.'* The analogy is a comparison device that establishes the similarities (and differences) between two or more things or situations. Analogies are a commonly used process to extend our thinking from the known to the unknown, allowing us to grasp new concepts or situations from familiar ground. Most of our reasoning and decision-making processes relate present decisions to similar experiences, and often when describing a new situation to listeners, we link it with something in their experience.

We thus use the analogy to describe the familiar aspects of a situation and identify differences as exceptions. The analogy is a familiar tool, particularly in the world beyond the scientific community, where distinctions are less precise and where judgment is exercised more intuitively. And nowhere is the notion of decision making more intuitive than in the local government environment where short- and long-term goals, conflicting constituencies, and financial responsibility must be balanced within the political process. Within this process, formal analogies can be used for both the predictive and the measurement tasks that are required to successfully assess the impact of a technology. What the analogy lacks in scientific rigor, it more than makes up in flexibility, comprehensiveness, and ability to measure external factors.

* T. A. O'Connor, "Methodology for Analogies," Office of Research Analysis, *Report* 70-0003, U.S. Air Force (June 1970). Quoted in Joseph P. Martino, *Technological Forecasting for Decision Making* (New York: Elsevier, 1975).

FORMAL ANALOGY

If an analogy is going to carefully compare the chances for success or the impacts of a given technology in one situation with other similar technologies or situations, it must take into account the following factors: technology, economics, management, politics, society, culture, intellectual standards, religion, ethics, and environment. Note that technology is but one item in an array of 10 factors that must be considered in constructing an analogy. Technology in its own right is critical, but it receives less weight in the analogy than in other previously mentioned processes.

An analogy is constructed to measure a new situation (or technology) against a previously experienced situation, establishing similarities and differences in order to obtain a better understanding of the new situation. If the similarities are great enough, then the outcome of the new situation should be similar to the outcome of the previous situation. If we are careful to establish an analogous situation, then we can make decisions with a reasonable degree of assurance that we understand the consequences of the decision. The analogy can be used as a predictive tool—that is, we can set up a hypothetical situation and analyze its consequences in light of our present information base in order to determine if a particular course of action should be pursued. It can also be used to assess the consequences of a decision already made or an action already taken. Whether the analogy is used in a predictive or an assessment sense, care must be taken to ensure that the two situations being compared really do resemble each other. This need to ensure that the situations do match point for point gives rise to the formal analogy, a process whereby two or more situations are measured from all points of view to ensure that they genuinely do correlate.

Even after point-for-point correlation has been ensured, the original situation must be carefully evaluated to be certain that the convergence of some unusual circumstances did not provide results that are historically unique. Any technology developed during World War II, for instance, would have to be carefully evaluated to ensure that technology needs generated by the war did not accelerate or decelerate its development so much that an analogy with present-day technology could not be created. With the length and intensity of U.S. resource commitment to that war effort, it is doubtful that any technology developed during that period was not hastened or delayed.

The procedure for constructing a formal analogy has five steps. First, evaluate the situation at hand to determine its major characteristics. Second, look for similar, but not necessarily identical, situations that are familiar or known well enough to be evaluated. These situations will form the basis for the analogy. Third, evaluate the situations chosen for comparison with the present situation, and ensure that there were no major forces that produced an unexpected set of results or could produce drastically different results from the present situation. The factors listed for a formal analogy—technology, economics, management, politics, society, culture, intellectual standards, religion, ethics, and environment—make good starting points for this analysis. Fourth, make a comparison of the two situations in each of the areas to ensure that they are truly analogous, that they do correlate. If there is substantial agreement, then inferences for a present situation can be drawn from the previous situation. If there are differences, they should be identified and categorized as significant or unimportant. A significant difference can make all or part of the analogy invalid, whereas a minor difference could have little impact. Fifth, if the situations are analogous, then the results of the past situation can be used to predict the probable outcome of the existing situation, with suitable allowances for minor differences.

ELEMENTS OF A FORMAL ANALOGY

In order to conduct a formal analogy, each of the elements must be examined to ensure correspondence between the two situations. Technology, economics, and environment represent the physical science elements and are thus measurable in objective terms. Management and politics represent the human element of the situation and must be treated with a sense of the here and now. Society, culture, intellectual standards, ethics, and religion represent measures of our civilization that are longer term and slow moving and must be treated with some historical perspective.

Technology

When constructing an analogy, three aspects of technology must be considered: competing technology, supporting technology, and complementary technology. For the basic assessment process, competing technologies must be looked at both for the existing case and for its complement on the other side of the analogy. It is necessary to look at both the primary technology and its competitors since a competitor could well supplant the favored technology. The amount of competition in each case can be important to the validity of the analogy.

The generation of electrical energy using nuclear power is one area of technology where, even at a national level, the choices among technologies are difficult to make, and the investments and consequences are substantial. Breeder reactors are capable of producing a greater amount of fissionable material than are conventional reactors and can therefore provide additional fuel supplies. In addition, the technology is closely allied to conventional fission reactor technology. There are some unknowns, however, and the energy-generation process still leaves us with radioactive spent fuels, which pose a storage and disposal problem.

Fusion by contrast uses common isotopes found in water, which are combined in a reaction to produce a harmless gaseous material, helium, along with energy. However, the fusion process involves extraordinary reaction temperatures and pressures, and we are not sure that the technology to produce controlled fusion reactions on a commercial scale can be successfully developed.

Perhaps both of these technologies will collapse from their own cost and complexity. Maybe a decentralized, nonpolluting hydrogen energy system will be the wave of the future. In making an assessment or forecast of electrical energy generation beyond the year 2000, there is no simple, clear-cut set of answers and more than one technology must be considered. In fact, our national agenda in energy reflects clear-cut bet hedging, with investment in each of the technologies, with the assumption that a clearer path will evolve in the 1990s.

A particular item of technology cannot be considered in isolation. The supporting technologies required to use it and keep it operable must be considered as well. Again, referring to nuclear power, our ability to use breeder reactors may be limited, not by the ability to design and build the facilities, but by our ability to safely dispose of the spent, still radioactive fuels until they have completely decayed and become inert. Even after the technology has been developed, citizens must be convinced that the technology is mature and therefore something they will tolerate. If both of these steps are not taken, further development of fission reactors for electrical power will come to a halt.

There is also a need to have a thorough fundamental understanding of the physical principles underlying each technology. In the case of a complex system, the information base includes a body of analytical knowledge supporting the system, or another system that is built into it. In any event, the technology must be considered in terms of all the elements required to allow it to stand alone.

In addition to the technology in question, complementary technology must also be considered. A particular technology is usually a part of a continuum of devices or processes that perform a function, produce something, or provide a service. Therefore, the technology must be considered initially by itself and then as part of the continuum. In the case of a transportation system, it may be something simple, such as compatibility with a standard track width or truck bed height. In data processing, common language and message transmission protocol would be a consideration, particularly as data communications become more common, interlinking previously isolated data processing machines. In a system that converts solid waste to energy, the performance of a waste to energy recovery device could be heavily dependent on the efficiency of a complementary piece of technology that recovers ferrous or nonferrous metals from the waste stream before it is converted to energy.

In a less direct realm, the analogy process also requires that we compare the level of understanding and development of the laws, theories, and explanations of each technology. If the laws and theories that underpin a technology are well understood, then the performance of the technology can be predicted with some degree of accuracy. If, on the other hand, the performance is based on sheer empiricism, with little understanding of the mechanisms of the technology's operation or performance, the ability to predict the pace and course of the technology is severely limited, and a valid analogy would be difficult to construct.

Economics

Technology is intended to serve a useful function. In order to do so, it must meet either of two economic criteria: It must provide a net return on investment, or it must accomplish a function at less cost.

In the first case, the implicit assumption is that other competing resource investments can be made. All organizations have competing priorities, and the events of the 1970s demonstrated that even the federal government does not have unlimited resources and fiscal pressure on our tax structures is building. And certainly all local governments are in the midst of difficult choices about priorities. Therefore, a primary measurement for the evaluation of a given technology that is competing with other technologies must be that it provides a greater return than its competitors. The rate of return must consider initial investment, continued operating cost, and eventual replacement cost. Although these initial capital, operating, and replacement costs can be measured in dollars, the nondollar resources required to support the investment must also be considered. These nonmonetary resources include personnel and facilities that are not directly attributable to a given technology but contribute to the infrastructure of the supporting organization. These support costs can and should be quantified when considering cost.

When a forecast or comparison is made, all costs should be considered. In the case of a telephone system, the initial focus is on the central switching system that connects the individual telephones to the larger telephone network. The switch system is the major capital investment in a purchased system, but not the major cost when a 20- to 30-year life-cycle cost is considered. With the relatively long life to be expected from telephone equipment, amortized equipment costs become small when compared to annual costs for personnel to operate the system and service contracts to provide maintenance. In this instance, far more research must be invested in the relative reliability of the equipment and the degree of automation that can significantly reduce the costs associated with personnel. Because costs represent both financial investment and resource investment, any comparison of the impact of competing technologies should look at the range of total resources required for successful implementation and operation.

Some benefits can be measured directly as returns on investment. It is easy to calculate the costs saved when more roads are paved per public works dollar spent or when more efficient collection equipment allows more refuse to be collected in an expanding community without increasing the work force. However, other benefits are nonmonetary and thus more difficult to measure. Providing the capacity to perform a task more thoroughly or accurately may be hard to qualify in economic terms, but it may still be a worthwhile goal. An information system that makes more detailed information available to citizens during the public planning process typifies this kind of activity. Some technology may provide no direct economic return, but it may provide social benefits that are greatly desired by a community. Many environmental improvements fall in this category. In these instances, some correlation must be made between economic investment and social rewards.

Very often there is no opportunity to measure the relative benefit of a particular service. Federal or state mandates or citizen initiatives may dictate a course of action, and the objective under those circumstances is to provide a maximum amount of service within a given budget or to minimize the cost of delivery. Under such circumstances, it is not a question of return on investment, but a matter of accomplishing a required function at a minimum cost.

Management

Two principal concerns are the breadth of the impact of a technology and the depth of skills necessary to support it. The size of the managerial task is largely measured by the number of people a particular task touches, either for the performance of a function or for the delivery of services. Two major not-for-profit firms, Battelle Columbus Laboratories and Public Technology Incorporated, recently conducted an experiment in technology transfer for the National Science Foundation to test the premise that innovations introduced to a local government by a full-time technology transfer specialist (a technology agent) could be transferred (replicated) to another local government with a minimum of intervention. The project was structured to be user oriented, and the six participating jurisdictions chose three innovations for transfer from a list of 28 that had been chosen by a research staff as having the highest potential for providing direct benefit to user cities and counties. The users chose telephone cost control, project control, and solar water heating as the three innovations.

Of the three, telephone cost control, a program for reducing costs of new or existing systems, was the least complex. It could be done with little or no capital investment and managed by a single person within a local government with passive cooperation from department heads and agencies. Project control, a formal scheduling and tracking system for capital investments and other major programs, called for little capital investment, but it did need the ongoing cooperation of all departments that were required to provide status reports and take corrective action to keep projects on schedule. Solar water heating required nominally $10,000 to $15,000 in capital investment, and it called for localities to go outside the government structure to obtain the services of professional architects for design and specifications. In addition, solar collectors and supporting equipment would have to be purchased from suppliers.

Not surprisingly, the ease of implementation was in direct proportion to the amount of managerial complexity involved in completing the project. It is worth noting that the requirement for capital and use of outside resources made the solar water heating project extremely difficult to implement. The amount of capital and the level of outside resources were not considered significant at the outset of the project. However, the need for appropriations from the governing body and for design service contracts significantly increased the number of people involved. The project changed from an internal activity to a task requiring public approval, and this considerably complicated the managerial process, decreasing the chances for success.

As already mentioned, the project control activity required little capital but continual ongoing resource commitment. This meant that a small portion of the resources required to support the project had to come from each participating department on an ongoing basis. This continuing resource commitment was a consideration in implementing a project control system, and in communities without a strong formal planning tradition that recognizes the returns to be had from a planning investment, project control is a difficult system to initiate. The telephone cost control project, on the other hand, can be either a single-shot investment or an ongoing process, depending on community needs and desires. In addition, its level of investment can be controlled, providing flexibility in an innovative project. One of the major results of the research was the recognition that complexity and extent of involvement should be major considerations in the implementation of innovations.

In addition to the breadth of impact, the depth of skills required to support an activity is an important consideration. It is extremely important at the outset to ensure that there is a matching of capabilities and technology when a new technology is being considered. Will the using organization understand the concepts behind the technology? Does it have the skills to operate and maintain it? Is the proposed technology evolutionary, or is it a step jump? If it is a step jump, then intensive training and a longer shakedown period may be required.

If new technology represents a major change, it must be carefully nurtured. The users may reject it because they are afraid of the changes it may cause or because of inertia or lack of desire to adapt to a new approach. However, the most difficult obstacle to overcome is overt opposition by people who fear new technology or do not understand it.

A major aerospace corporation, turning its acquired skills to the municipal–commercial sector, developed a device to control the flow of water from fire hoses in fire fighting operations. The unit used a radio transmitter built into the hose nozzle to permit the fire fighter using the hose to remotely control water flow by operating a flow control valve located in the fire pumper. The unit automatically adjusted the flow and pressure in response to the fire fighter's signals, eliminating the requirement to adjust pressure manually at the pumper control panel, and represented a major step forward in fire fighting technology.

Fire fighters, however, despite their equipment orientation, are remarkably resistant to change. They distrusted the complexity of the equipment and by and large would not accept it. In addition, the innovation was promoted in some municipalities as a labor-saving device, since it automated a time-consuming manual function. These two factors, technical complexity and potential for crew size reduction, in conjunction with some early electronics reliability problems were enough to doom the product. The fire fighters resisted it, and eventually it was removed from the market. In terms of straight technology, it presented a significant advancement in fire fighting. In terms of the other factors surrounding it, it must be measured today as unsuitable technology for all but a few jurisdictions that can resolve the management problems associated with its use.

Politics

The larger the departure from accepted practices a technology causes, the greater the potential for change. It is a truism in engineering that any change causes subsidiary ripple changes. Some of the ripples can be anticipated, some cannot, and the human interface is difficult to quantify. Since politics is the art of balancing competing interests, the major questions related to the introduction of new technology are who benefits and who loses.

The previously described system to control water flow from fire hoses provides a good example of how the political strength of an organization can affect the introduction of new technology. From a manager's point of view, the flow control system could automate the manual pressure-setting function at the pumper, thus either decreasing crew size or making more people available for actual fire combat. The net result is more available people at the fire scene per dollar expended. From the fire fighter's point of view, the flow control system must be backed up by a person monitoring the pumper controls, and any attempt to reduce crew is looked upon as a dollar-saving gambit that endangers the safety of the fire fighters and the public they are protecting. The question of acceptance in this efficiency versus safety trade-off is largely determined by the political strength of the fire fighters, and in most communities, this strength is considerable.

Some municipal departments, such as the police department, that perform a regulatory or prohibitive function have a negative public image that largely limits their political power. Fire fighters also exercise a public safety function, but one that is viewed more positively. They perform a rescue function, saving lives and salvaging property. As a result, there is considerable political support for their activities, and fire fighters have become quite adept at wielding this power effectively. Thus, the technological innovation—the automated flow control system that is often opposed as a labor-saving device by fire services—gets tied up in the emotional question of saving lives and becomes a liability if a mayor, manager, or fire chief is forced to test his influence against the rank-and-file fire fighters in the public arena.

In the assessment process, the relative strength of those who will benefit from the innovation, and will thus support it, must be assessed. By the same token, who loses and how much they lose must also be evaluated to ensure that an innovation will have enough support to succeed. In this instance, perceived losses by some party are just as crucial as actual losses, except that improper perceptions can sometimes be changed by education and discussion. Some fire services have enthusiastically embraced the water flow control system, since they perceive it as a way to provide better services at lower cost. Obviously, the acceptance process is much influenced by the traditions, attitudes, and perceptions of a relatively influential local government service organization.

Society and Culture

Technology must be considered in the light of whether it aids or conflicts with basic societal or cultural values. Cultural values are the sum total of our individual backgrounds, education, goals, hopes, and fears, and they reflect the relative order of things good and things bad. In short, our cultural values reflect our perceptions of what is good and what is bad in our society. Major institutions and organizations serve as vehicles to express the social and cultural values we have formed to work and live collectively. Major societal institutions include governments, schools, businesses, and families.

There is no doubt that the technological orientation of our culture permitted a massive investment of resources and skills to land a man on the moon in the late 1960s. The voyage to the moon was made possible by the efforts of a technically trained population that was capable of developing the hardware and systems and by strong popular support for the national goal of lunar exploration. The nation responded positively to an initiative put forth by a popular president, the representative of all society's hopes and aspirations.

The same public, on the other hand, caused the resignation of a president in 1974 when he violated a basic cultural value. In this instance, the key societal institution, the Congress, moved very slowly, and ultimately public opinion, expressing our cultural values, deposed the president for breach of honesty.

Traditionally, institutions have had a major impact on our ability to accept and integrate new technology. The construction industry, made up of small companies and dominated by strong labor unions, has long resisted time- and labor-saving techniques because of their impact on total employment within the industry. Dock workers fought containerization of ship freight for similar reasons, and the railroad unions have institutionalized featherbedding in freight and passenger activities.

Technology usually changes more rapidly than institutional

structure, and industries that are impeded by institutional structure either lose ground to competing technologies or are forced to
move in new directions in order to survive. In the long run, these
structures are overcome, but in the short run, they can be powerful
forces that significantly influence the course of technology.

The institutional response of industries also determines the
relative acceptance of technology. Basic industries, such as steel
and heavy manufacturing, which traditionally do not make major
investments in research, find themselves in trouble with competing companies from Europe and Asia that have modern plants
and more modern production techniques. The United States
aerospace and electronics industries, which traditionally invest
heavily in research and development, have remained competitive
with foreign industry and are faring well domestically and in the
international marketplace.

Local governments have an institutional response to technology
that is rooted in their service delivery orientation. Since local
governments are the front line of service delivery, citizen attitudes and perceptions will have much to do with the acceptance
or rejection of an innovation. Here the major consideration is
whether the change will be internal or external. An internal
change that is not highly visible can be easily accommodated. On
the other hand, technology that makes a visible change in service
delivery, or that is subjected to the public deliberation process,
must be carefully measured to ensure that it is deemed to be
consistent with our societal and cultural values.

Intellectual Standards

Intellectual standards are those established by leading individuals and institutions who are considered to be opinion formers by
our society. Some of these opinion formers may be instrumental in
establishing new trends; others may be the protectors of time-
honored traditions and standards. They have in common a widespread acceptance as advocates of the common good and thus
command considerable influence within our society.

The technology transfer programs that were started in the late
1960s and early 1970s all called for a healthy involvement on the

part of opinion makers, because they reflected the prevailing intellectual standard that technology could and should be used to solve some of our societal problems. As the programs evolved, however, this stance of influencing intellectual standards deteriorated to one of pandering with elected officials, governments, and institutions for additional funds for individual programs.

There are some supportable reasons for a shift in emphasis: a need to publicize success to justify funding, some well-thought-out plans to expand the scope of technology involvement, and local government's desires for federal assistance with pressing problems. There are also some unsupportable reasons: bureaucracy building, individual profiteering, and a tendency to adopt the classic approach of drowning problems in money as a substitute for careful management and hard work.

In any event, many of the better technology transfer programs quickly lost their intellectual appeal and foundered when they became submerged in politics.

Religion and Ethics

Technology that is perceived to be in conflict with religious and ethical values will have little chance to succeed. When technology is measured against religious and ethical standards, it is measured against internal concepts of right and wrong and not against externally imposed standards. A major example of technology being in conflict with religious standards is the controversy surrounding abortion. A better understanding of physiology, advances in sterilization, and special-purpose surgical equipment have made abortion a potentially routine surgical procedure. However, significant opposition by citizens whose religious principles argue against taking human life has established strong sentiment against abortion, particularly where communal (public) rather than individual funds are used.

The science of genetic engineering is also receiving considerable attention. As our knowledge advances, the concept of creating or substantially altering life forms comes closer to reality, and there is much fear that this technology will be misused. The future path of genetic engineering will be heavily influenced by our religious and ethical beliefs about the sanctity of human life, and regulatory boards and commissions have already been established to ensure that progress in this science remains inside established ethical boundaries.

Our investment in nuclear power is also bounded by the knowledge that widespread nuclear production and development capability brings with it the specter of misuse for destructive purposes and that the peaceful uses must be limited to prevent widespread adoption of atomic weapons for warfare.

However, all of this does not mean that religion or ethics and technology do not mix. Christian TV networks are an outstanding example of the use of modern technology to spread the word of God, and television ministries have been used extensively to generate contributions to support church-related activities.

Environment

In the first half of the century, maximum attention was paid to industrial development and little attention to the ability of our natural systems—air, water, and land—to adapt to the changes industry has imposed on them. In the second half of the century, we reaped the results of our previous efforts—smog, dirty rivers, and diminishing scenic and open areas. This led to a series of initiatives at the federal level to force the benefits of technology to be measured against the consequences to our natural systems. Regulations concerning air, land, and water are now firmly embedded in our structure of national laws. As our knowledge of consequences and harmful effects increases, our ability to measure costs as well as benefits of new technology gets better, and environmental costs become a greater consideration. This is particularly true in such areas as mining and chemicals, where costs of by-product disposal are now being allocated to the production process.

In our effort to modify the world energy situation, decreasing our dependence on Middle Eastern oil by shifting to domestically available substitute and synthetic fuels, there are substantial environmental trade-offs to be made. As we learn more about the damaging effects of production and use of fuels, the equations relating environment, health, economic, and political stability become more complex and more difficult to balance effectively.

SUMMARY

In constructing a formal analogy, all factors must be considered, both in absolute terms for the situation at hand and in relative terms when the present situation is compared with a similar situation either to predict a future course of events or to measure the consequences of a decision to adopt or not to adopt a particular technology. The analogy is a powerful tool for analyzing technology and for measuring its fit with a given situation. Care must be taken, however, to assure that the two situations being considered are genuinely comparable if valid results are to be expected.

CHAPTER 7

Models

A model, as defined by the Oxford English Dictionary, is "something that accurately resembles something else" or "a thing that represents on a small scale the structure of something greater." As the dictionary definition indicates, a model is something that simulates or represents the performance of a real system. The value of a model lies in its ability to economically predict the future outcome of a set of actions or the impact of changes to a system. The model accomplishes this function by simulating or imitating the performance of a large-scale system and producing approximate results in an abbreviated time span. Time and labor are saved by abbreviating the time span; simulating the performance of actual systems is accomplished by using mathematical equations to describe their response. Particularly where the operations of equipment and machinery are simulated, modeling can produce orders-of-magnitude savings in costs.

There are some limitations, however. The word *approximate* is important. Scaling down any dimension, time included, is bound to introduce some distortion. In addition, a model imitates a real system, and it does so accurately only to the extent that the model builder is able to comprehend all the elements of a real system. The fidelity of the model is thus limited by the modeler's willingness and ability to reduce an actual system to mathematical terms.

In the long run, the only system that produces valid results is the real system, but modeling can be a time- and resource-saving activity that has real potential in the evaluation and decision-making process. Martino, in his book *Technological Forecasting for Decision Making,** states that "the current state of the art of technological forecasting models is quite primitive." His statement can be applied to many of the tools currently available to simulate our social and physical systems. In present modeling practice, much emphasis is placed on curve-fitting techniques to make sense of scattered data and on the development of precise prediction curves based on a few historical variables.

Local governments are complex social–physical systems that are difficult to imitate with simplistic prediction or assessment tools. When predicting or assessing technology in the context of this complex system, the technological component occupies only a small place in the total realm of social, political, economic, ethical, and aesthetic considerations. In light of this, any tool for assessing consequences or predicting a path must move away from elaborate mathematical characterizations based on a few easily measured variables and move toward accounting for the many contributing forces that are working in concert with, or in opposition to, one another. More variables complicate the calculation process, producing many more options to evaluate, and the process of building the model and evaluating its results becomes very complex.

Given this pessimistic vision of where we are today, there are many things happening in the development of models that provide some optimism for the future. The advent of the digital computer, a complementary technology, has made complex modeling a reality. The computer has the ability to store large amounts of information, perform lengthy calculations, and manipulate and summarize data. It can also make decisions in accordance with preprogrammed instructions. These capabilities can reduce weeks of calculations to minutes. The computer can also manage a wider variety of options than can be accommodated by the intuitive thought process, by virtue of its ability to make decisions and choose alternatives according to established rules at any point

* Joseph P. Martino, *Technological Forecasting for Decision Making* (New York: Elsevier, 1975).

during the calculation process. In parallel with computer development, we have evolved mathematical descriptions of our physical and social systems and predictive techniques that make large-scale modeling possible. Every election day showcases the progress that has been made in forecasting. Election results are routinely predicted with 2, 5, or 10 percent of the voting count in with surprising accuracy.

Despite the sometimes impressive examples, modeling as a science has a long way to go. Much progress has occurred in the physical sciences, where rules of physics, chemistry, and mechanics are straightforward and easy to interpret. Social systems, by contrast, have many more inputs and reactions and are more difficult to categorize and describe. For instance, in the last several years, major efforts have been made at a federal level to understand the impacts of increasing reliance on foreign oil and of OPEC's pricing policies. By and large, we have not been able to track our energy future with any accuracy. Petroleum prices have risen far beyond projected levels; we have had economic disruptions but have not suffered catastrophic dislocations; and the decrease in energy consumption predicted much earlier is just now starting to take effect. Even in a typical, narrowly isolated case—the prediction of price elasticity of electrical power consumption (the effect electric rates have on per capita consumption)—it has been difficult to provide simple answers over the last several years because of the complex and intertwined issues of energy surcharges, energy conservation, changing rate structures, and the controversy over nuclear power safety and reliability. Although there may be long-term correlations, the present data scatter is far too great to permit drawing conclusions.

Despite the complications, our body of knowledge is growing and with it the ability to simulate the performance of our physical and social systems.

TYPES OF MODELS

There are five types of models available for use in local government activities. Probably the best established and most widely used are population projection models. Physical system simulation models, such as those used to evaluate storm water drainage systems, water supply, wastewater treatment, and refuse collection and disposal, are seeing wider use. Service delivery models that can be used to locate emergency service centers, such as fire stations, have been developed but are not widely used at present. The increasingly tight financial environment in local governments has given rise to financial forecasting models that can not only predict the income implications of community growth but focus on the cost consequences as well. Special tools are also available to provide formal structure to the decision-making process. Finally, all these elements can be interconnected to provide an integrated, flexible, comprehensive planning process.

Demographic Models

Models that predict population trends are the most familiar and commonly used local government modeling tools. Demographic models use data on birth rates and death rates in conjunction with immigration and outmigration data to forecast changes in population. Demographic data provide the basis for projected changes in facilities, services, and, indirectly, the tax base.

Decision-Assisting Models

Battelle Columbus Laboratories has developed the technique of "interpretive structural modeling" as an aid to decision making. The model organizes and formats complex issues so that they can be categorized or clustered as related activities and ranked in order of importance. It introduces objectivity and a certain amount of formality into the process of decision making or priority setting, thereby avoiding some of the digressions and disagreements that usually accompany a broad agenda-setting process. It also reduces polarity by separating issues and forcing a simple value judgment on related activities.

To start an interpretive structural modeling activity, all the issues associated with the activity in question—health care needs, for instance—are listed with the objective of prioritizing and categorizing needs so that goals can be established and resources allocated in accordance with the goals. Interpretive structural modeling adds structure to the decision-making process by creating relational statements. A typical statement is: "Issue A is more (less) important than Issue B." The decision-making group discusses the statement and arrives at a yes or no decision by majority vote. The results are entered into a computer that serves as a scorekeeper. The group then moves on to the statement "Issue B is more (less) important than issue C" and again discusses the statement and arrives at a group decision. The results are entered again in the computer, and the process continues until each of the issues in the sequence has been ranked against another issue.

As a result of the first round of comparisons, some relationships have been established. For instance, if A is more important than B, and B is more important than C, then A is more important than C, and that question does not have to be asked of the decision-making group. In similar fashion, if E is related to F, and F is related to G, then E is related to G, and that question does not have to be asked. For the next round the computer automatically establishes these relationships, cataloging the ones that have been determined and isolating the comparisons that have not been made. The computer then suggests pairs of issues to be compared for the next round. The next round then proceeds. For instance, B and C may have been rated of equal importance, leaving the relationship between A and D undetermined. The statement "Issue A is more (less) important than issue D" would be suggested by the computer for discussion.

The process proceeds, with fewer questions to be discussed at each succeeding round, until all relationships have been ranked. The computer is then able to print out the issues in rank order. If the issues are particularly far ranging and diverse (when trying to establish overall departmental goals, for instance), an additional statement, such as "A is (is not) related to B," is used along with the ranking statement. This serves to establish groups of related issues and permits ordered priorities within the groups.

This separation and grouping process can be particularly helpful when the decision-making group contains strong advocates for different areas. The process forces the group to consider sets of issues separately and sidesteps the possibility of deadlock over unrelated issues. Each issue is discussed within the context of its own group rather than in competition with other sets of issues. When all the relationships have been established, the model produces ordered rankings of issues and, if appropriate, groupings of issues for further discussion and refinement.

Interpretive structural modeling, like most modeling tools, serves as an aid to decision making. It shapes and organizes the process so that the major issues can be clearly prioritized and subjected to further study and scrutiny. In recognition of its role as an aid to the decision-making process and not a decision-making tool itself, the computer-ordered results are carefully scrutinized after apparent closure has been reached to ensure that none of the inferred conclusions are irrational. At any time during the process, any relationship can be redefined in accordance with the participants' intuitive logic, and the process can be run until closure is reached again.

Interpretive structural modeling is an excellent vehicle for organizing and sorting issues at the beginning of a program or during the public participation phase where a range of issues must be confronted and priorities must be established.

Physical System Simulation Models

Physical system models simulate the performance of actual systems by using a series of generalized equations to define the response, in time and in space, of the physical process to be modeled. The equations are derived from clearly defined physical laws or from theory verified by actual experience. In the case of processes that are difficult to define, empirical relationships are sometimes established. This may also be done when the body of knowledge is largely empirical, coming from observation of results rather than analysis of operation.

The generalized equations in the model are then refined by actual data about the characteristics of the particular system to be modeled, tailoring the model to the actual system being simulated. Design stimuli—input information in such standard formats as a design rainfall in a hydrological model or traffic density in a traffic model—are applied to the model, producing output information about the performance of the system. In actual practice, easy-to-understand historical inputs are run through the model to verify that the model produces results similar to those actually experienced. This process verifies that the model does simulate the real system. In some cases, when model and actual results are divergent, equations are modified to produce better results. Once the model has been calibrated for the system in question, it is ready for use.

A hydrological model is a typical tool for simulating the performance of a physical system. It contains complex generalized mathematical equations that describe what happens to rainwater as it strikes the surface of the earth and flows in time and space through a watershed to its ultimate destination. In so doing, it characterizes the drainage patterns of the watershed, in particular for seldom occurring but devastating heavy rains. The equations are based on physical laws of gravity, hydrology, and fluid flow, supplemented with empirical results obtained over years of specific experimentation. Essentially, the laws predict how the water will travel in time and space from the time that it rains until the time the water reaches the river, stream, or outfall that is the outer boundary of the watershed with which we are concerned.

Shape, slope, and vegetation characteristics of each subregion in a watershed are inputs to the model, along with descriptive data about major streams and channels, in order to make the model specific to the area in question. More detailed models characterize natural impoundments (lakes, ponds, and storage basins) and such man-made features as drainage pipes, channels, and conduits. The end result is a mathematical description of the drainage paths in the area in question and the way that water flows through them.

Design rainfalls, either typical or worst-case rainstorms, are inputs to the model. The outputs are hydrographs—curves showing the flow of water over time at crucial predesignated points in the watershed in question. The model pinpoints potential flood conditions and provides an indication of excess stream and channel capacity. It can also be used to model improved drainage structures, impoundments, and response to catastrophic rainstorm.

Standard engineering equations create a snapshot of performance for a given situation at a given time. The equations are good, simple tools for producing average design conditions and some maximum and minimum results, but they are incapable of accommodating complex inputs, particularly when they take place over time. This snapshot calculation does not account for transient situations, where things are changing rapidly. The model can repeat the equations over and over again in infinitesimal time increments, producing time-related results. This is particularly important in drainage, where the cost of designing systems that will accommodate a storm that occurs once every 50 or 100 years far exceeds the benefits and some compromises must be made. A model can simulate the consequences of such a storm, where small changes in the system can drastically alter flood patterns.

The model is particularly useful for designing impoundments that store water to avoid peak flows or for designing channels, conduits, and pipes to retard flow. Such techniques reduce flooding downstream in a watershed by slowing flows and spreading them over a longer time period. If the impoundment and retardation devices are not properly designed, however, they can produce worse conditions than free-flowing systems since stored water can be released suddenly. The model can simulate worst-case situations to ensure that system failure does not occur.

Water quantity models have matured to the point where they can be expected to produce predictable results and are finding their way into drainage sections and public works departments in cities and counties around the country.

Water quality models, on the other hand, are less mature than water quantity models. Quality models are a direct outgrowth of the quantity models and use quantity model data and equations to predict surface water runoff and stream and river pollutant flow. Equations for the generation and absorption of pollutants are superimposed onto the water quantity data in order to obtain predictions of pollutant levels in streams and rivers. The water quality management activities sponsored by the EPA—the so-called Section 208 plans—have made water quality a necessary ingredient in the resource planning arena.

The planning process works at determining pollution sources and assimilative capacity of receiving streams in order to set the stage for pollution abatement in an area. Sources can be point (pollutants that can be clearly traced to a single source, such as a factory or wastewater treatment plant) or nonpoint (agricultural runoff in rural areas, fertilizer runoff in suburban areas, and street debris in urban areas, for instance). Assimilative capacity refers to the ability of a waterway to cleanse itself—to remove pollutants by natural processes as the water flows.

The area-wide Section 208 management plans, in their attempts to strike a balance between generation and capacity, have relied heavily on modeling techniques to predict pollutant loads. The dynamics of river assimilation capacity, however, are much more poorly understood than the dynamics of water flow. In addition, data on nonpoint pollutants and generation rates are largely empirical, and this body of knowledge has not been fully developed. Because of these limitations, water quality modeling has lagged behind quantity modeling and is still in its infancy.

Municipal water distribution and wastewater collection models have performed a function similar to that of the drainage system models. The major difference is that water distribution and wastewater collection systems are not created by nature. As a result, they follow simple laws of physics and hydraulics and require little empirical data for the production of good results. The major problems with these man-made systems relate to the number of control actions that can be taken to modify pressures and flows within the system. Characterizing the operation of valves, pumps, and numerous other control devices is a major challenge in modeling man-made systems.

A water distribution system model simulates the hydraulic response of a system as water flows from the supply source to the ultimate users. The distribution network is described by entering information about tunnel and pipe lengths, interconnections and elevations, and friction losses in pipes, valves, and fittings. The operation of pumps, regulators, and control valves that appear in the system is also described in mathematical terms. Daily demand curves simulating groups of individual, domestic, and commercial water requirements are applied to the distribution nodes (end points). The model works its way back up to the supply point, calculating pressures and flows at critical points in the distribution system along the way. Supply system characteristics as well as tank storage and pump operations can be simulated.

The wastewater collection system model operates in much the same fashion as the water distribution model, except that the wastewater input into distributed nodes is simulated and the flow is traced as it moves from the source toward a central collection point at a wastewater treatment plant. Many wastewater collection systems rely on gravity flow for simplicity, and therefore terrain is important, as it is in the hydrological model. Some generalized wastewater transport models are capable of planning the routing of a pipeline according to terrain. They are also capable of sizing and costing pipe collection networks, using standard installation costs for piping. The terrain (pipe slope) and flow rate determine pipe size and land configuration determines length, so that approximate costs for systems can be developed. Wastewater quantity and composition are determined by using typical flows and waste concentrations for residences and by using actual or empirical numbers for industrial and commercial waste.

Treatment as well as collection processes can be modeled. I have developed a design model for characterizing the performance of municipal solid waste incinerators. Figure 12 shows a block diagram of the model. Garbage quantity in tons per hour and energy value in BTU per pound of garbage are inputs to the model. The furnace section of the model simulates the combustion process where the garbage burns by itself with the addition of air.

The model calculates the amount of ash generated as residue and the amount of particulate carried away from the burning grates, using empirical data. A chemical composition for the refuse is assumed, for purposes of calculating the quantity of flue gas from the combustion process, as well as its temperature and moisture content. The flue gas is then carried through a spray cooling system or a steam-generation section, where its temperature is reduced. As the gas passes through the furnace breeching on the way to an exhaust fan, its temperature drops further and it loses some pressure. The last step in the process is an electrostatic

FIGURE 12. Model for municipal solid waste incineration.

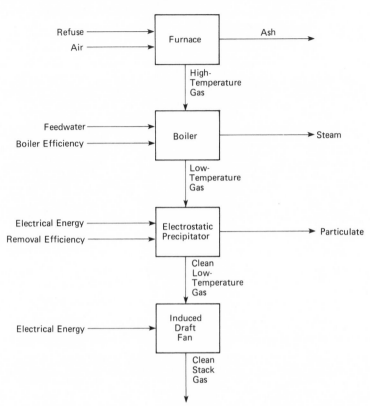

precipitator element that simulates the particulate removal process, using an equation that calculates exit flue gas particulate concentration as a function of inlet velocity and particulate concentration of the incoming gas.

The addition of composition data (percentages of ferrous and nonferrous metals, glass, combustibles, and noncombustibles) and removal processes, as indicated in Figure 13, converts the model from an incineration process model to a resource recovery process model. Figure 13 shows a refuse-derived fuel process that has provisions for shredding, inert material separation, and ferrous metal removal. With the model, and some data on equipment efficiency and operating costs, the performance and operating economics of a resource recovery system can be calculated, taking

FIGURE 13. Model for resource recovery process.

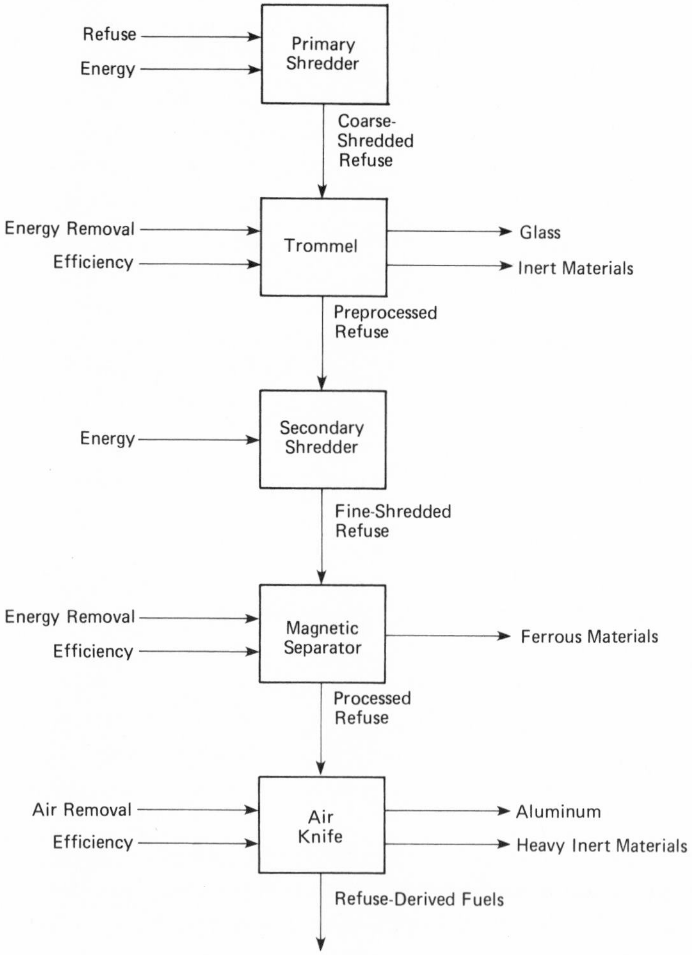

into account the considerable variability inherent in a solid waste stream.

Service Delivery Models

Service delivery requirements can be modeled by using methodology similar to that used for physical system models. At least one organization has developed a fire station location model that is the first in a series of potential service delivery optimization models. The fire station location model develops centroids of potential fire demand within a locality and calculates the optimum mix of fire station locations to serve the community efficiently. Potential fire demand is determined by evaluating the nature of the building uses in a given zone (residential, commercial, industrial) and assigning a desired response time in accordance with the mix of buildings. The model calculates the travel time from the existing (or proposed) fire station to the demand zone, correlates zones with stations, and shows where response does not meet specified levels. Alternative combinations of stations and response times can be simulated until a balance between resources and risks has been arrived at. The fire station location package can be and has been used to plan new station locations, exchange or eliminate existing station locations, and evaluate response time against preestablished standards.

The fire station location model calculates transit time from a central support facility—the fire station—to specific service points. If this process is reversed, the service points become points of demand and the central support facility becomes the center of a service demand zone. The location model can then be used to determine optimum locations for service delivery centers, such as health clinics, mental health centers, and libraries, to name a few. The model would consider travel times for individual and mass transit for determining service center locations. Some inferences would have to be made as to what segments of the population will be heavy users of a given service and where they are located, but actual service data will provide a strong starting point. The same process can be used for establishing satellite offices or combined service centers for all services—tax collection, licenses, inspections, and so on—in order to provide convenient local service delivery.

Financial Models

In the financial area, several models have been developed to calculate the cost of new development in a community. The primary revenues associated with new growth are fairly easy to predict. They include fees from the planning and construction process and additional property taxes from new construction. Derivative income–sales taxes from additional economic activity, additional employment, and commerce created by additional population are somewhat more difficult to trace. The costs of growth, however, have been largely ignored because of a fundamental belief that growth is universally beneficial. With the advent of decreased population and economic growth, the era of growth for the sake of growth is apparently nearing an end. Movements for restricted growth or no growth tied with a desire for better quality of life are manifesting themselves around the country. In addition, inflation has eaten heavily into the traditional slow-moving property tax base of most local governments. All these factors point to a more careful assessment of growth in the future and more attention as to what kind of growth makes a net positive contribution to a community.

Several financial models have been constructed to formalize the process of calculating the costs of growth. Major activities affected by increased growth are physical facilities, such as roads and utilities, and basic mandated service or protection functions, such as fire, police, and schools. Other human services may briefly lag behind basic services, but equity considerations and citizen demand will require that all services be provided to new areas or communities as well as to older sections.

Facility-oriented services require capital for equipment and some personnel costs for maintenance and operation. Human and protective service functions are labor intensive but require capital for equipment and facilities. All functions require expenditures for shelter, energy, communications, and supplies as well as for such additional support services as personnel, record keeping, and supervision.

Supplying personnel for support services can often be characterized as a continuous function. Teachers can be added to the work force in proportion to population, as can fire fighters and police officers. However, facilities must be added to support and house these service delivery personnel. These additions occur as discrete functions. Teachers are added when space permits (and sometimes when space does not permit) until facilities are saturated. At some point, a decision is made to add more facilities. This step jump in physical capacity requires two to five years to accomplish and brings the personnel–facility ratio back to a reasonable norm. This process can be modeled by using standard ratios of service delivery personnel to population and adding facility costs in discrete steps corresponding to increases in service personnel. The model thus relates service delivery costs to service delivery requirements.

The expenditure side of the financial model can mesh the incremental needs for people with the discrete capital requirements for facilities to evaluate the time-related aspects of providing support services and, to a certain extent, to permit some growth optimization that can be tied to a locality's ability to sustain additional general fund and capital fund pressure.

A complete financial model compares projected income from primary and secondary sources with the flow of expenditures required to support additional development so that decisions about growth can support overall community goals. In areas where population is declining, it can be used as the basis for the allocation of shrinking resources or for the development of a business retention strategy.

DATA BASES

The physical system and service delivery models previously discussed are driven in part by natural features and in part by the actions of the community they describe. With the exception of the solid waste model, which simulates a specific process, the models also have a strong geographic spatial distribution component. The input information—water consumption for the water distribution model, wastewater generation for the wastewater model, land changes for the hydrological model—are population sensitive. Population and employment center information can be translated to input data by using conversion factors. The information can be gathered manually or input automatically by an information system or data base that records business locations and population characteristics.

One such data base, which is sponsored by the Census Bureau, is the DIME (Dual Independent Map Encoding) file. The DIME file has been offered to municipalities in major metropolitan areas as a means of making census information more available for analysis.

Basically, the DIME file characterizes a community through its street and road network. A not-to-scale map depicting all streets as lines and all intersections as nodes is created, and each street and node is given an identifying numerical label for use by the computer. Each street-side (block) address range is identified and attached to the nearest street node to create a file of street addresses and nodes. Data gathered during the census are added to the file by matching street addresses listed in the DIME file. The resulting product is a computerized file of demographic data that can be identified at a block level in accordance with Census Bureau regulations. Individual family and personal data are not provided but are aggregated to a block level. By adding geographic or political boundaries to the file, census data can be gathered at any desired level—traffic zone, assessment district, school district, election district, city, county, or watershed.

The DIME file has some shortcomings. In its present form, it is spatially but not geographically oriented. It does not contain dimensional data. If geographic coordinates were added to each street node, the DIME file could become a geographically based system that could be linked to the physical system and service delivery models to provide the input information that would drive them.

There are other data base systems in varying stages of maturity that subdivide geographic areas into discrete parcels that contain geographic coordinate and terrain elevation data. Demographic data are then added to the coordinate data to provide a complete data base system.

Census data provide a basis for modeling existing and short-term needs. However, if the physical models can be appended to a periodically updated predictive or projective tool, such as the land use plan, then a dynamic predictive process can evolve. The land use plan, which ideally reflects citizen desires for the future shape of a community, serves as a guide to elected officials in the day-to-day zoning process. By using this plan as a driver for the physical system models, the impact of future changes on existing or projected systems can be evaluated. If a shortage of capability in some physical system manifests itself, it can be evaluated by making simulated changes in the physical system or land use plan or both. Any number of proposed changes can be iterated back into the model, allowing a range of workable alternatives to be developed and evaluated in terms of economic, social, political, and other costs. The value of the process is that it gives decision makers some insight into the consequences of their decisions and at the same time provides a series of alternative decisions or modifications that can help maximize the desired benefits and minimize the undesirable consequences. Figure 14 shows the relationship of each of the models to the overall decision-making process.

There is always some danger that the data developed by the models will develop an aura of authority that precludes further changes to the process. This can be overcome by recognizing at the outset that models at best can only mimic a real system and thus only serve as indicators of consequences. If this approach is taken, then model data will be only one part of a continuous process of planning, setting goals, implementing, measuring progress toward goals, and then repeating the process. This is particularly applicable to the continuous process of charting and framing a locality's growth and change.

SUMMARY

Models have the capability to plot the cumulative effect of individual decisions. The decisions about land use and development are made on a week-by-week or month-by-month basis and are frequently isolated from one another. Models have the capability to determine the total impact of individual decisions. Models also provide a quick assessment of alternative courses of action. Multiple alternatives can be evaluated from a single input without

FIGURE 14. Modeling the decision-making process in an urban system.

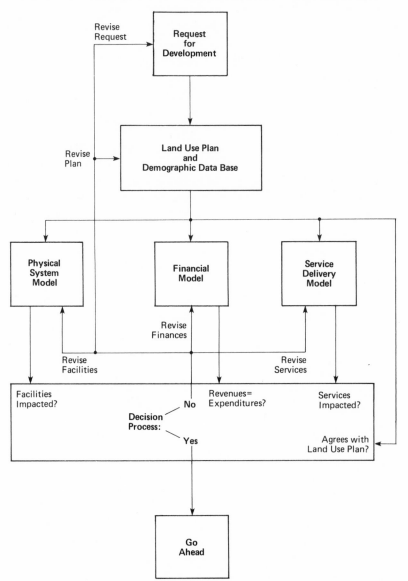

the time-consuming manual task of adding up individual costs and benefits.

Models also allow combinations of strategies to be evaluated. In physical systems with large capital investments, various strategies to maximize the use of existing facilities can be tried and the best combination of existing utilization and new investment can be developed. Policy alternatives can also be evaluated. For instance, decreasing or increasing service delivery levels can be readily measured in terms of impact on facilities and personnel.

Transient problems can also be modeled. In drainage modeling, for instance, water storage capacity can be substituted for pipeline or channel capacity, decreasing the need for larger pipes or channels by spreading the drainage of water accumulated from rainfall over a longer time period. Similarly, the impact on schools of a temporary increase in children concentrated in a given age group (a wave effect) can be accounted for.

A good model will be structured to provide alternatives that will allow elected officials to weigh objective consequences against such subjective information as voter attitudes, economic climate, and other competing priorities in order to provide a decision best suited for a locality's constituents. Properly done, the modeling process will increase rather than decrease alternatives by providing more specific cost and benefit data about options.

The value of models lies not so much in their ability to provide an exact picture of where we will or should be, but in their ability to allow alternative paths, readjustments, and deviations while we are still pursuing a set of general goals.

CHAPTER 8

Driving Forces in the 1980s — The Environment

The 1970s were years of massive environmental legislative activity at the federal level. Citizen fervor about the environment, which gained momentum in the 1960s, is largely considered to have peaked on Earth Day in 1970. This fervor led to new legislation and major additions to existing legislation that established goals for the quality of the environment and set limitations on what could be done with our air, land, and water. This legislative work, which was largely completed in 1978, established the framework for the 1980s.

On January 1, 1970, the National Environmental Policy Act became law. It established a council on environmental quality in the executive office of the president and required federal agencies to prepare environmental impact statements on proposed legislation and "other major federal actions significantly affecting the quality of the human environment." In the same year, the 1970 Clean Air Act was enacted, revising the Clear Air Act of 1963. This act directed the establishment of national air quality standards, and extended emission standards to mobile sources as well as to stationary sources covered in the 1963 act.

In 1972, Public Law 92-500 made far-reaching changes to the

existing Federal Water Pollution Control Act. The Marine Protection Research and Sanctuaries Act (Ocean Dumping Act) prohibits ocean dumping for some substances and regulates the dumping of others. In the same year, the Federal Insecticide, Fungicide and Rodenticide Act (FIFRA) of 1947 was amended, creating the Federal Environmental Pesticide Control Act, which significantly increased federal regulation of these substances. The Noise Control Act, which regulates mobile and stationary noise sources, also became law in 1972.

The Safe Drinking Water Act, which set standards for public drinking water supplies, was passed in 1974. In 1976, the Toxic Substances Control Act and the Resource Conservation and Recovery Act (RCRA) became law. The Toxic Substances Control Act regulates toxic materials not controlled under previous acts, while the RCRA establishes requirements for the disposal of solid waste and requires a permitting program for the generation, transportation, and disposal of hazardous wastes.

All these acts are administered under the aegis of the United States Environmental Protection Agency, which was established in 1970 to administer federal environmental programs. The acts administered by the agency differ widely in their impact: The Toxic Substances Control Act and the Pesticide Control Act apply primarily to industrial manufacturers and suppliers; the Water Pollution Control Act, the Clean Air Act, and the Resource Conservation and Recovery Act apply both to industry and to local governments; the Safe Drinking Water Act affects municipal water suppliers; the Ocean Dumping Act applies to coastal areas; and the National Environmental Policy Act, while aimed at federal activities, filters down to all levels of government through the federal grant process. The laws also differ widely in their approach and implementation. The National Environmental Policy Act states sweeping policy goals. The Water Pollution Control Act establishes water quality goals, sets discharge standards, and provides municipalities with grant monies to construct facilities that will meet the discharge standards. The Safe Drinking Water Act also sets standards, in this instance for water supply, but provides no money for constructing or modifying facilities to meet the standards.

These acts contain two common provisions that apply to all municipalities. The first is that any citizen can instigate a civil suit to force compliance with the provisions of the laws. The second is that governments that do not comply with the provisions of the legislation face a cutoff of federal funds for other programs.

This chapter will outline the major provisions of each piece of environmental legislation that impacts on local governments and will attempt to trace the present and future consequences of each of the acts.

THE NATIONAL ENVIRONMENTAL POLICY ACT

The National Environmental Policy Act requires the development of an environmental impact statement (EIS) for any proposed legislation or activity that has a significant impact. Although the requirement applies specifically to federal agencies, it extends to activities that are carried on by local governments and funded wholly or partially by the federal government. Because of the pervasiveness of federal funding, and various permit requirements, almost any major facility now constructed requires an environmental impact statement. Such statements tend to be voluminous and are expensive to prepare, and there is some question about why a separate policy act is needed when the other federal environmental acts require the evaluation of activities that they govern. However, the EIS does provide a formal review of the environmental consequences of any major activity—something that had been too often lacking in the past.

THE WATER POLLUTION CONTROL ACT

The Water Pollution Control act and the Clean Air Act have the most far-reaching effect of any environmental legislation to date. The original comprehensive legislation, the Federal Water Pollution Control Act of 1956, empowered the federal government to make studies of water quality in interstate water bodies. If the body was polluted, the federal government took steps to eliminate the discharges causing the pollution through a conference, hearing, or court suit procedure. The law was further amended in 1961 to include all navigable waters.

By 1965 many conferences and hearings had been held, but none of the open cases had been resolved. Congress, therefore, amended the law to establish standards for interstate streams. States were asked to establish standards for the segments of the stream in their jurisdiction, and the EPA was empowered to set standards if states did not submit their own standards or submitted unacceptable standards. The intent of the amended law was that water quality below the accepted standards would be evidence that dischargers were violating the law. The dischargers could thus be taken to court to be made to comply. This strategy was challenged in the courts since it was difficult to attribute poor water quality to any individual polluter among a large group in a waterway.

In 1970 the federal government discovered a provision in an 1899 act that prohibited discharges into navigable waterways without a permit issued by the U.S. Army Corps of Engineers. Despite the fact that the law was an amendment to the Navigation Act and not an environmental law, it was a useful tool and was invoked by the government. Dischargers again sued, contending that the 1899 act was not an environmental act, and the stage was set for the sweeping changes encompassed in Public Law 92-500, the Federal Water Pollution Control Act of 1972.

The 1972 act establishes national goals for clean waters. It requires that dischargers obtain permits; it establishes abatement requirements for industry; it requires a minimum of secondary treatment for municipal wastewater treatment plants and makes up to 75 percent funding available for design and construction purposes. The act further requires the establishment of stream standards and an area-wide waste treatment management plan. It also establishes research and development grants, and authorizes the EPA to establish regulations governing the cleanup of oil and other hazardous substances from navigable waters, along with innumerable other activities.

The law has two goals that are often erroneously interpreted by the public as regulations:

It is the national goal that the discharges of pollutants into the navigable waters be eliminated by 1985.

It is the national goal that whenever attainable, an interim goal of water quality which provides for the protection and propagation of fish, shellfish and wildlife and provides for recreation in and on the water be achieved by July 1, 1983.

These "zero discharge" and "fishable–swimmable" goals are separate and distinct from national policy, as stated in the law, and further yet removed from the effluent limitations and stream standards that the EPA is authorized to establish. They were, and still are, very ambitious targets that provide focus for national pollution control efforts, but they do not fall into the category of enforceable law.

Section 301 of the law requires states to establish water quality standards for all interstate waters that are water quality limited and not presently meeting national goals. The states are required to establish effluent limitations as well as a continuing planning and prioritization basis for the allocation of federal monies to upgrade wastewater treatment plants for those streams already receiving too much effluent.

Section 208 establishes the concept of controlling point and nonpoint sources of pollution. As mentioned earlier, point sources of effluent are clearly identifiable sources from a public or private wastewater treatment plant or an industrial process plant. Nonpoint sources are diffuse runoffs from the land. In an urban area, contaminants would include storm water runoff from streets and developed areas containing litter, rubber, asphalt, dirt, and other contaminants. In a rural area, contaminants could be soil from erosion and farming, pesticides and insecticides, and waste matter from animals. In strip mining areas, contaminants could be acid runoff from mine tailings.

Point source effluents, although expensive to clean up, are fairly simple to categorize, and equipment is available to significantly reduce discharges. On the other hand, nonpoint sources by their nature require more diverse and indirect control measures, and the impact of these effluents on the ability of a water body to clean itself up and sustain aquatic life is not yet clearly understood.

The area-wide management process for wastewater treatment must be marked as a failure at this point. Part of the failure is due to the late start the EPA made in implementing the management process. It followed, rather than preceded, the establishment of water quality standards as required by Section 301 of the law. Since municipal treatment plant construction priorities were already established, the area-wide management process could not significantly influence the ongoing design and construction programs. In addition, there is much work yet to be done in the area of stream capacity and water quality measurement. Although the water sampling programs built into the management studies are a good starting point, they have not produced enough information to characterize the waste-carrying capacities of water bodies. Given the lack of good data, few communities and industries are willing to reduce their effluent levels to meet standards that are not well defined and documented.

The most awesome provision of the management process, however, is the concept of establishing land use controls and development limitations in order to reduce the point and nonpoint effluents to the level required to attain good water quality. This section of the law was widely viewed by localities as an attempt by the federal government to establish land use controls, a function that has been historically performed by the localities themselves. Since most water bodies are shared by localities, few were willing to apportion effluents among themselves, and fewer still were willing to make the social or capital investments the management plans would have dictated. Given the political, economic, and social consequences, it is doubtful that the area-wide management programs will succeed.

The law also establishes effluent limitations for private and publicly owned treatment works. It requires private treatment plants to use "best available technology" (BAT) by July 1, 1983, to limit effluent discharges. The BAT requirements have not been met, and there is some question as to whether the investment to apply BAT is reasonable. In light of these findings, BAT provisions were pushed back as much as four years in 1977 legislation.

The general requirement for publicly owned treatment works is that they use secondary wastewater treatment that reduces biological oxygen demand (required to decompose organic materials in the wastewater) by 90 percent, reduces the suspended solids (particles of waste matter carried by the water) by 90 percent, and kills bacteria and viruses by disinfection of the wastewater (usually with chlorine) at the end of the treatment process. In cases where streams are water quality limited, advanced wastewater treatment is required. Advanced treatment removes plant nutrients, such as phosphorus and nitrogen, and further reduces biological oxygen demand and suspended solids.

The law provides for payment of up to 75 percent of the cost for design and construction of a new or upgraded plant, but it does not provide for maintenance or operations money. As a result, most treatment plant designs emphasize low capital cost at the expense of operations and maintenance capability. As might be expected, treatment costs rise asymptotically as effluent cleanup gets closer to 100 percent, and advanced wastewater treatment plants are very expensive to build and operate.

As of 1977, only 40 percent of the major municipal dischargers were in compliance with the law.* Construction activity has lagged far behind congressional authorizations and indeed far behind EPA appropriations. This lag can be attributed partly to a refusal by the Nixon administration to allow congressionally authorized funding levels to be reached in the early 1970s and partly to the long, complicated planning and authorization process that even those municipalities willing and eager to improve their wastewater treatment capacity are forced to endure.

Combined sewer overflows are a particularly costly problem for older municipalities. In these cities, the runoff from storm water is conveyed through the same collection and treatment system as the wastewater. Rainfall causes an enormous surge in the distribution system, requiring much of the storm water and wastewater to bypass treatment plants. This adds substantial contamination to receiving waters during periods of rainfall.

The solutions to the problem of combined sewer overflows are not easy or cheap. One alternative, separating storm water and

* Council on Environmental Quality, "Water Quality," *Environmental Quality—1979,* Tenth Annual Report (Washington, D.C.: Government Printing Office).

wastewater flow, is an overwhelming task in most cities and economically impractical. Another alternative, designing treatment plants to handle maximum combined flows, has some major disadvantages. First, it is economically unreasonable to build and operate a plant whose normal utilization will be less than half of its capacity for almost all its operating life. Second, biological treatment plants are incapable of responding to the rapid step increases and decreases in wastewater flow caused by rainfall. A third alternative, using a combination of storage capacity combined with somewhat larger wastewater treatment capacity to level out the surges caused by rainfall, provides a good, if expensive, compromise solution and one that is being pursued in some areas of the country.

Despite substantial construction activity in the 1970s, on the part of both municipalities and private industry, water quality has on the whole just about held its own. Increasingly now, nonpoint pollution sources are being recognized as major contributors to water pollution. As more and more industrial and municipal wastewater treatment plants come on line, contributions from nonpoint sources will become a larger portion of the wastewater introduced to our streams and rivers.

In general, our efforts at water pollution control have reached some level of maturity. The major facets of the problem are amenable to a structural solution—the construction of wastewater treatment plants by both the private and public sectors. Private sector compliance is reasonably good. The publicly owned treatment plants are lagging, but federal funding for construction is significant, and we are turning the corner on public treatment plan completion. There is some question about whether new plants can meet treatment standards and whether we are maintaining existing plants adequately. However, these problems are controllable. Nonpoint pollution sources are looming larger as a problem. We do not presently understand enough about their impact on streams, or even about the ability of waterways to clean themselves up. Furthermore, nonpoint sources do not lend themselves to convenient structural solutions. They will have to be attacked through other approaches, some of which will be opposed by local governments and individual citizens.

There is no doubt that the national goal of zero discharges will not be met by 1985, and there is some question as to whether it ever can or will be met. The most rational solution appears to be a better understanding of the assimilative (waste absorption) capacity of our receiving streams, tailoring effluent outputs so that the assimilative capacity is not exceeded. That level of knowledge, however, lies in the future.

THE CLEAN AIR ACT

The Clean Air Act has two major aspects. First, it imposes individual limits on the stationary and mobile sources of pollution. Second, it establishes standards for the quality of the air itself in terms of allowable contaminants. The act thus controls major sources of air contamination and establishes further requirements for allowable air quality by state-determined regions.

The act authorizes the administrator of the Environmental Protection Agency to establish primary and secondary standards for the following five basic pollutants: particulate, sulfur oxides, carbon monoxides, nitrogen dioxide, and hydrocarbons. Primary standards are those required to protect human health. Secondary standards are intended to protect public welfare and the quality of life and are more stringent than primary standards. The responsibility for measuring and enforcing air quality rests with the states.

In 1970 amendments to the Clean Air Act required states to submit state implementation plans (SIPs), which would spell out in specific detail how federal goals would be met. The original date of July 1, 1979, for submission and approval by EPA was met by only one state—Wyoming. Forty-four states submitted SIPs by the end of July, but few plans have been fully approved, and in March 1980, the federal government moved to cut off federal aid in the state of Colorado for failure to make a good faith effort in its air quality planning.

The Clean Air Act of 1977 revised earlier deadlines for actually attaining ambient air standards. The target date for complying with standards for particulates, sulfur dioxide, and nitrogen dioxide is 1982; it can be as late as 1987 for ozone and carbon monoxide in some circumstances. Given the slow pace in the submission of state implementation plans, it is doubtful that the target dates for complying with the standards will be met.

The SIPs have been slow in coming because the states are required to institute several controversial measures and to implement some soft strategies that have far-reaching consequences.

Areas within a state that meet or exceed air quality standards (attainment areas) must be designated as Class I, Class II, or Class III under the nondegradation provision of the Clean Air Act. This provision, established as a result of a suit by the Sierra Club, is a compromise between the lofty goal of nondegradation and a firm policy of attainment of national secondary standards.

Class I is an area that includes major recreation, scenic, or historic areas where any air quality deterioration would be considered significant and therefore not permitted. Class II is an area where the deterioration in air quality that accompanies moderate growth would be acceptable. Class III is an area where intensive industrial development, permitting air quality to deteriorate to secondary standards, would be permitted. In no cases would deterioration be allowed to go below secondary standards.

The EPA at the outset designated all areas except national parks as Class II and has allowed the states to re-classify areas as Class I or Class III. In addition, new emission sources in Classes II and III attainment areas are required to use best available technology, regardless of cost, to minimize emissions. In nonattainment areas, states are required to consider a variety of tools to bring pollution levels back to secondary standards. Some of the more controversial approaches include automobile inspection and maintenance, emission offsets, and transportation and parking control programs.

Auto emission programs are controversial because of their impact on the consumer and because of questions about their effectiveness in assuring compliance with emission standards. Existing state automobile inspection programs concentrate on auto safety and can be implemented with existing skills. They vary in effectiveness from state to state, and there are real questions about how to run such a program effectively. Emission inspection programs require a significant step-up in knowledge and equipment in order to be effective, and there is some question as to whether they can be successfully built into existing safety inspection programs. Nevertheless, proper auto emission system maintenance heavily influences emissions, and an inspection program counts heavily in any proposed emission reduction program. In cases where the ozone attainment date is extended from 1982 to 1987, Congress has mandated the auto emissions inspection program.

Emission offsets are required of any new polluting facility in a nonattainment area. Any new facility must use best available technology for emissions and, in addition, must offset its emissions by using the emissions allocated to a previously existing facility as long as the combined emissions total less than previous emissions. This offset can be obtained by closing an existing facility or by installing new emission control equipment. It essentially amounts to the purchase or auctioning of emission capacity. As a business interchange process, it can be workable; when industry and local governments get involved in industrial development, it will become a delicate and complex process.

In recognition of the economic consequences of limiting effluents, the EPA has instituted a "bubble concept," whereby facilities will be allowed to reduce emissions in an area or process that is easier to upgrade, and therefore less costly, to offset increased emissions in a more difficult and costly area, provided that overall facility emissions do not exceed allowable standards.

Transportation controls present another formidable problem. Central cities, already plagued with economic problems and anxious to retain business and employment, will be extremely reluctant to limit parking and ban vehicular traffic. Although mass transit is envisioned as a way to replace individual vehicular traffic for the most crowded (and therefore most polluted) areas, there is going to be a significant response gap. Today we do not have even the capacity to build sufficient buses to expand our transit system, much less the capital and institutional arrangements that will be required for an orderly transition to mass transportation. Recent Reagan administration initiatives to reduce federal funding for mass transit programs will place additional pressure on cities and states.

In addition to establishing overall air quality standards, the Clean Air Act has established requirements for stationary and vehicular emission sources. The stationary source emission requirements have been in place since the 1970 act, and in general, there has been considerable movement to ensure compliance. However, there has been some question about the EPA's technology-based standards that specify the appropriate control technology. These standards are still being published for some industry segments, and there are inefficiencies built into their proscriptive nature. Among other things, it is argued the standards are not the most cost-effective approach and they discourage innovation. By specifying technology, the EPA removes some of the compliance responsibility from the polluters.

The use of wet scrubbers to reduce sulfur oxide emissions is a case in point. The EPA has specified that electric utilities use wet scrubbers in plants that burn high-sulfur coal. The scrubbers use limestone or similar material mixed with water to clean up stack gases. The mixture is sprayed into a chamber, where it comes into contact with the hot flue gases. The water entrains particles, while the limestone or other agent reacts with the sulfur oxides to remove them from the gas stream. The process is elaborate and expensive, and produces a sulfur-bearing sludge that ultimately is landfilled. The utilities have fought this cumbersome, expensive process, and in the long run, they will not be entirely responsible for the poor results from it, since it was specified as a solution.

Regulations affecting mobile emissions are changing the shape of the automobile industry. The regulations for carbon monoxide, unburned hydrocarbons, and mixed oxides of nitrogen are specified on a per-vehicle-mile-driven basis (grams per mile). This absolute limit encourages the use of smaller engines, since such engines can emit larger percentages of pollutants in their exhaust gases and still meet emission requirements with simpler emission control systems.

Emission controls are also pushing gasoline engines into a much narrower operating band, in some cases decreasing drivability. The requirement to reduce unburned hydrocarbons and carbon monoxide emissions moves in the direction of increasing combustion efficiency and hence overall engine performance.

However, as combustion efficiency increases and the fuel and air are more perfectly matched and burned, combustion temperatures rise. These higher temperatures in turn cause increased production of mixed oxides of nitrogen (NOX). Thus, in order to decrease NOX emissions, combustion temperatures must be dropped below what would ordinarily be produced by the ideal conditions that decrease carbon monoxide and unburned hydrocarbon emissions. When these conditions are complicated by the requirement to maintain them during start-up, warm-up, and acceleration in addition to during normal cruising operations, the operating regime of the internal combustion engine becomes very inflexible and difficult to manage. As the emission requirements stiffen, the acceptable operating band gets narrower and more elaborate controls are required to maintain standards. More complex controls mean more complex adjustment procedures, more maintenance problems, and, to the average motorist, more operating headaches.

The diesel engine, which offers greater economy and lower pollutant emissions, is in limbo because of concern over potentially carcinogenic emissions and will likely be in partial production until lab testing is completed and emission standards are finalized. Other engines that offer promise, such as external combustion engines and gas turbines, have not produced anticipated fuel efficiencies and require further development.

The automakers themselves are experiencing the severe trauma of transition from a free market industry to one that is heavily subject to federal regulation in the areas of emissions and product safety. In 1980, the trend to smaller automobiles and engines, which American automobile manufacturers had been resisting, was given sharp impetus by a drastic shift in consumer preference. This trend, which had been gradual, was greatly accelerated by the advent of dollar-per-gallon prices for gasoline. Consumer preferences may change overnight; highly automated mass production assembly lines require years and billions of dollars to convert. The Chrysler Corporation has been brought to the brink of disaster by the twin forces of regulation and consumer taste shifts and requires federal support to avoid insolvency. The Ford Motor Company has also suffered huge losses because it retained its emphasis on larger cars. Only the General Motors Corporation, the largest of the major auto manufacturers and the one with the most assets, has positioned itself to meet changing buyer preferences.

THE RESOURCE CONSERVATION AND RECOVERY ACT

The Resource Conservation and Recovery Act (RCRA) of 1976 is one of the newest large-scale pieces of environmental regulation, and its effects have not yet been felt. As a new law, it completes the water–air–land cycle of total environmental regulations. It establishes requirements for the closing of open solid waste dumps and requires the administrator of the EPA to establish standards for the design and operation of sanitary landfills. It encourages resource recovery by the establishment of regional technical assistance panels and the financing of resource recovery feasibility studies. In the area of hazardous wastes, the EPA is required to develop a list of hazardous wastes, establish requirements for the safe disposal of these wastes, and develop a womb-to-tomb manifest system that will trace hazardous wastes from the time of generation through ultimate disposal. As with most of the other laws, authority for development of plans and regulation of wastes is delegated to the states, with provisions for assumption of responsibility by the federal government in the event of a default.

The major weakness of the law is that it does not address one of the major issues that will confront us in the 1980s—the emergency cleanup and the continued maintenance of closed hazardous waste dumps. As the experience at Niagara Falls, where homes were built adjacent to an abandoned hazardous waste landfill, clearly indicates, significant health hazards can derive from long-term exposure to hazardous wastes. In an effort to resolve this problem when existing owners do not have the resources to clean up hazardous waste sites, or in cases where ownership has changed, the Congress has passed legislation authorizing a "superfund," to be funded partially by a tax on chemical companies and other producers of hazardous waste. The superfund would be available to provide for emergency cleanup of hazardous waste sites, and it would be maintained by a combination of taxes, government appropriations, and monies recovered from polluters after emergency cleanup has been accomplished.

The act is moving toward implementation, but, because of the complexities involved, it is much behind schedule. The solid waste portions of the act, although behind schedule, are moving slowly toward fruition. The hazardous waste portion has not had the benefit of several years of research as has the solid waste area, and the three-part activity—identifying waste, establishing a permit process, and defining requirements for hazardous waste disposal facilities—is tied up in research requirements and crisis cleanup situations.

Solid Waste

In April 1980, the EPA promulgated some thousand pages of regulations for disposal of solid waste in landfills. Although the regulations have not been implemented, they will surely have significant impact on ground water contamination, disease reduction, and decreased litter and air pollution. They will also double or triple existing landfill costs by establishing stringent design, operation, and postclosing maintenance requirements.

In recognition of the complex technical, legal, financial, and institutional problems associated with the development of an efficient and environmentally sound solid waste collection and disposal system, the EPA established several levels of technical support activity. Although these activities focus on all aspects of collection and disposal, they emphasize resource recovery as a desirable national goal because of its positive economic, environmental, and energy impacts.

At the first level, the EPA provides technical seminars and peer match programs for interested local officials. The seminars, usually two-day technical sessions, are aimed at familiarizing local officials and others in the solid waste industry with the latest solid waste collection and disposal techniques. The peer match program is aimed at municipalities and brings together local officials who want to improve their solid waste situations with officials who have direct experience in upgrading their own operations.

At the next level, the EPA provides professional assistance to municipalities in each EPA region through technical assistance panels contracted for by the EPA and staffed with experienced solid waste professionals. Although the panels are capable of providing technical support for any collection or disposal problem, they are generally employed to provide municipalities with a strong conceptual starting point for a complex and demanding process, such as the establishment of a resource recovery facility.

At the highest level of resource commitment, the EPA, through the President's Urban Policy Program, has awarded grants to 68 municipalities throughout the United States to proceed toward the implementation of resource recovery facilities. These grants differ from wastewater facilities grants in that they fund only feasibility studies and do not pay the actual construction costs of facilities. In solid waste, the trend has been to make resource recovery facilities self-sustaining through a combination of charges to facility users (usually municipal or contract waste haulers who pay "tipping fees") and sales of materials or energy. At present materials and energy prices, resource recovery progress costs cannot be fully recovered without a tipping fee, but this activity is an extension of an almost universal practice of charging service fees based on waste weight or volume to cover amortization and operating costs at private and municipal landfills.

There have been attempts to establish resource recovery facilities (primarily steam-generating incinerators) in the United States over the last 20 years, but the results have been mixed at best. Several steam-generating municipal incinerators are successfully operating, but not selling steam. A pioneering effort by the Union Electric Company in St. Louis to burn solid waste along with coal foundered because of a state law prohibiting utilities from recovering capital until a facility was actually generating power and because of citizen opposition to waste collection site locations. Well-conceived facilities in Baltimore, Maryland, and New Orleans, Louisiana, must be rated partial to total failures. The resource recovery industry is risky, and the risk is aggravated by a lack of ongoing projects that would provide designers, builders, and operators with a continuing history to build on and evolve from. The sustained level of federal wastewater treatment funds provides the treatment plant industry with exactly this kind of advantage.

The problems of technical evolution and continuity pale, however, when they are ranked with institutional problems. The problems of getting municipalities to work together—establishing control of the waste stream, developing self-sustaining bonding and financial provisions, establishing acceptable user charges, signing long-term contracts with industry for energy and materials, and selecting the proper resource recovery ownership and operating options—overwhelm the technical issues. Activity, therefore, focuses on the institutional issues. In the long run, if they can be resolved, the technical and financial issues will fit into place. Although it is too early in the feasibility study process to determine the ultimate outcome in the 68 grant jurisdictions, the program can be expected to produce significant results.

The Department of Energy has a parallel program aimed at energy recovery that complements EPA's program, and the two agencies are coordinating their efforts in a concentrated attack on solid waste disposal.

Hazardous Waste

The hazardous waste program, which will have the biggest impact, particularly on the industries, over the next several years is just getting under way. Regulations for the permit (manifest) system for the transport of hazardous wastes have been published; some classification of hazardous waste has been started; and the regulations for hazardous waste disposal sites are in progress. These requirements must be implemented by the states, and it is anticipated that it will be 5 to 10 years before the regulations are substantially implemented. The first step, voluntary registration of disposal sites by industry, took place in the fall of 1980.

Hazardous waste activity will be substantial in coming years. The Council on Environmental Quality estimates the cost of cleaning up the many abandoned existing hazardous waste dumps at $28 billion to $55 billion, and the EPA estimates an annual cost of $750 million for hazardous waste management.

THE SAFE DRINKING WATER ACT

The Safe Drinking Water Act has many parallels to the Water Pollution Control Act: States are vested with the responsibility for establishing the program, water quality standards are established, and it authorizes continued research in improved water quality. The principal difference is that it requires compliance with standards without providing funds to assist in meeting the standards. This process has led to confrontation in the administration of a standard for one contaminant.

The law applies to public and private treatment works that support either 15 or more connections or 25 or more persons. It establishes primary regulations for the protection of health and secondary standards relative to drinking water taste, appearance, and odor. The primary standards set maximum levels for bacteria, turbidity, and organic and inorganic chemicals. There is also a requirement for sampling to ensure adherence to standards and for public notice (publication) if satisfactory results are not obtained. Secondary standards are not mandatory under the federal regulations but may be implemented by the states. The law also provides for state regulation of deep injection of water to prevent contamination of ground water aquifers.

The Clean Water Act reached the height of controversy in February 1978, when the EPA proposed regulations that would mandate the installation of granulated activated carbon (GAC) filters for all surface water supply systems that support more than 75,000 people. The requirement resulted from evidence that the chlorination and return of wastewater produced carcinogens—trihalomethanes—which showed up in downstream drinking water. The suspicion was buttressed by the widely questioned results of tests on the Mississippi River at New Orleans.

When the GAC requirement was promulgated, it triggered an immediate response from wastewater treatment plant operators, who considered the filters to be marginally beneficial and unnecessarily expensive to install and operate. The American Water Works Association (AWWA) protested that no water system that uses GAC treatment can meet the requirements of the proposed regulation. The Environmental Defense Fund (EDF), on the other hand, noted that if the GAC requirement is confined to cities with over 75,000 people only 52 percent of the present U.S. population would be covered. Faced with pressure from both sides, the EPA is assessing the need for changes before issuing final regulations.

SUMMARY

The wastewater cleanup activity governed by the Water Pollution Control Act is fairly mature. Industry has achieved about 75 percent compliance with the use of best practicable technology. The municipal picture is somewhat less encouraging. Although the EPA has obligated $20.7 billion for some 16,000 construction grants, only 5,776 grants encompassing $1.7 billion in funds were completed. Part of this can be attributed to the stop-and-go nature of the funding (federal funding was again withheld in mid-1980 to give the appearance of balancing the budget); part can be attributed to the complexity of the design and funding process; and part to the reluctance of municipalities themselves. The inflation rates of the early 1980s can be expected to hinder projects even further as spiraling costs cause refinancing and reprioritization of grant funds. But in general, the grant program, even if well behind schedule, is in place and operative and will start to produce significant results in the near future, even though results will be slowed by proposed reductions in future funding.

There are indications that even a complete updating of industrial and municipal wastewater plants will not be sufficient to attain fishable–swimmable goals. We will certainly miss the 1985 zero discharge goal, and there is good reason to question whether it should ever be attained.

As the effluents from point discharge are eliminated, nonpoint sources loom larger and larger in the cleanup picture. These sources, however, do not lend themselves to simple structural solutions, and the implications of such solutions—control of development and land use—will be accepted reluctantly at best by local governments. Much information must still be gathered about non-point contaminants and the impact on stream assimilative capacity, and it will be at least the end of the 1980s before we clearly understand the best compromise to be afforded between the present state of affairs and the 1985 goal of zero discharges. Even the direct structural measures to minimize point source pollution have been expensive, and the nonstructural methods will prove even more costly. Some planning will continue under the auspices of the Section 208 area-wide management program, but little will be done until a data base that leads to reasonable compromise has been developed.

The requirements for stationary and mobile sources of emission will be complied with by industry. The auto industry, pushed by a sudden consumer demand for small cars, will reach significant compliance by the mid to late 1980s; aggregate fleet mileage will rise and emission requirements will be easier to meet with a smaller average fleet size. If the carcinogen cloud surrounding diesels is lifted, there will be significant expansion in passenger car and light diesel use; diesels are almost universal in large equipment and over-the-road vehicles. Outside of diesels and derivatives of the Otto cycle internal combustion engine, little new production engine technology is anticipated for the 1980s. With the changeover to smaller engines meeting new emission standards and the phaseout of older passenger cars in the fleet, significant ambient air quality improvements can be anticipated. Maintaining these gains, however, will require steps on the part of manufacturers and owners to ensure that emission systems continue to operate within their design limits.

Congress has mandated new stationary source standards that require the removal of 70 percent to 90 percent of the sulfur dioxide emissions in an attempt to prevent some industries from substituting low-sulfur, western coal for high-sulfur coal instead of installing emission control facilities. Industries have expressed opposition to across-the-board scrubbing, and negotiated delays in the 1983–1984 implementation time will permit the development of promising alternatives to wet scrubbing.

The area of most difficulty will be state and municipal actions to ensure that ambient air quality standards are met. States have been moving slowly on implementation plans, partially because they have to set up an organizational structure and then draw up plans that meet federal requirements, but primarily because the planning requirements touch on issues that are difficult to decide on and implement. States are reluctant to pass regulations for auto emission testing because of the difficulty in administration and the individual cost of compliance that will be passed on to motorists whose vehicles will have to pass emission tests. Traffic control measures, which will have to be implemented by units of local government, will be particularly difficult. They are required in noncompliance areas but cannot be directly implemented by the states. Responsibility must be vested within individual localities, which will be reluctant to implement measures that would adversely affect economic growth.

Cities and counties will also experience difficulty brokering emission offsets but will be forced to assist in this process in order to continue to attract industrial investment. Although the bubble concept, whereby a single factory determines how to allocate its total emissions, will be a straightforward regulating and monitoring process, brokering emission offsets to introduce new industries will be much more difficult. The existing industries may not be willing to surrender even their unused emissions and may be concerned with the labor supply and wage pressures new industry could create. Local governments will thus be saddled with a sharing problem, and they will have little other than persuasion or coercion to assist in the situation unless they are willing to dig into the tax or utility base in order to provide incentives for one or both sides. Unless private industries agree to share emissions among themselves, the offset concept will be barely workable at a local level.

The Resource Conservation and Recovery Act, which closes the air–water–land environmental triangle, is just starting to take effect. If it is implemented as legislated, it will have a significant effect on the way the country handles its solid and hazardous waste.

To date, resource recovery processes have not succeeded on a large scale in this country because they have not been able to compete with landfilling as an alternative. There has been a history of incineration, a process that reduces solid waste volume significantly, but incineration has been employed primarily where population density and land demand have made landfills economically or socially unacceptable. Incinerators have high capital and operating costs compared to landfills. The emission limitations applied to stationary combustion sources in the 1960s added significantly to the capital and operating costs and the sophistication of incinerators, causing many existing units to close and halting the construction of new units. Heat recovery in the form of steam generation is an integral part of almost any new incinerator, but the addition of a boiler and operating controls again raises the capital cost of incineration and drives the operation closer to a utility power generation function and further away from a disposal situation.

The requirements of the RCRA will significantly increase the costs of landfilling. Open dumps will be a thing of the past, and new landfills will have to be designed to retain contaminants rather than passing the environmental cost on to the nearest aquifer or stream. The concept of internalizing the cost of proper landfilling will drastically alter the landfill–resource recovery relationship. The combination of higher landfill costs and higher energy costs coupled with potential oil and electrical energy scarcity will make more resource recovery systems, particularly energy recovery systems, feasible.

However, the federal government still has a significant role to play. New institutional arrangements based on a self-sustaining combination of energy and material revenues will have to be developed. Regional approaches that take advantage of the economics of scale will have to be developed, and in areas of lower population density, smaller-scale, lower-technology systems will have to be developed. Finally, some sort of continuous funding support, analogous to but not as extensive as the municipal wastewater treatment program, will have to be developed in order to convert from the present risky, one-at-a-time, stop-and-go ventures to a continuous stream of projects. This will allow design and construction skills to be sharpened and permit experience to overcome potential construction and operating problems. The resource recovery industry needs a history of continuous operation to get itself firmly established. The most likely impetus, if it comes, will be from private industry, taking advantage of accelerated depreciation and tax credits now allowed by law.

Hazardous waste will be a difficult problem for the 1980s. The hazardous waste manifest system will be cumbersome and time consuming, but it is expected to be operational in the next couple of years. Disposal facilities, however, will continue to be a problem. Large financial outlays will be required for emergency activities at existing sites, and continuing activity will be required to bring them into conformance. Those sites that are abandoned, or that have owners without substantial assets, will be a particular problem, since the state and federal governments will have to assume responsibility for them in order to protect public health and safety.

Locating sites for hazardous waste disposal will be even more difficult than it is for municipal solid waste disposal. There has been ample evidence in the last several years that municipalities are not willing to accept these facilities. This lack of acceptance may provide further impetus to build facilities that process hazardous waste for reuse either as feedstock or as fuel to deliver energy. Facilities that process industrial waste will become much more viable as disposal costs rise and fuel prices drive industry toward waste energy.

There is a major negative impact from the hazardous waste provisions. Marginal operations that produce hazardous sludges—waste oil refining facilities, for example—will be driven out of business by a lack of disposal sites and by the high costs associated with trucking and ultimate disposal of hazardous wastes.

Water suppliers will continue to vigorously resist new regulations, even if proposed standards are verified by testing and experience, until EPA regulations provide for partial reimbursement of the costs associated with the new standards. Water systems, whether privately or municipally owned, are customarily operated as self-sustaining enterprises. Such initiatives as the granulated activated carbon filter to decrease trihalomethane concentrations in supply water can significantly alter the revenue and bonding requirements for water supply systems.

The major environmental concern for the 1980s will be to maintain the gains made in the 1970s in the face of enormous pressure to develop domestic energy. These domestic alternatives—coal, shale, synfuels, and biomass—are available but will carry environmental penalties in their development, production, and use. This conflict is further aggravated by existing and proposed cuts in federal spending. The gains of the 1970s will not be lost, but there is some real question as to whether they will be extended.

CHAPTER 9

Driving Forces in the 1980s—Energy

The Arab oil embargo of 1973 produced a major change in the world energy situation. Oil, which had been a cheap commodity, suddenly became expensive. This major shift in price, imposed by OPEC, caused severe economic dislocations and massive redistribution of wealth. A period of major instability has ensued, as the developed countries attempt to adapt to the rising energy costs driven by petroleum prices and to respond to the crisis with alternate, less politically fragile energy sources.

How did we reach this point? Where do we go from here? This chapter will attempt to answer these questions within the context of an unstable, rapidly changing energy situation whose outcome can only be hinted at.

It took the United States more than 5 years to respond as a nation to the original oil embargo. Depending on world political response, stability can return in as little as 5 years, when the effects of incentives to accelerate exploration for additional supplies and the development of facilities for additional production are fully felt, or it could take as much as 20 to 25 years with a concerted program for practically eliminating dependence on foreign oil and gas supplies.

EARLY HISTORY

In the 1800s, our energy supply was largely derived from wood and coal. Oil was developed as a lubricant and artificial light source. In the early 1870s and 1880s, gas derived from coal was used for lighting. In the early 1900s, during World War I, oil was used principally for industrial and residential heating. Around the time of World War I, the Automotive Age started and with it the growth in the use of oil. In 1900, there were only 8,000 registered motor vehicles. This number grew to over 1 million by 1913, 10 million in 1922, 27.5 million in 1940, until by 1980 the number exceeded 132 million.

Oil production followed a parallel course. American oil production rose from 64 million barrels in 1900 to 1.4 billion barrels in 1940. In 1982, domestic production of 10 billion barrels has been supplemented by 8 billion barrels of imported oil. Oil use continued to climb in other sectors as well, and by 1950, it had replaced coal as the predominant energy source in this country.[*]

Natural gas demand followed a similar course. Originally a discarded by-product of oil extraction, it was turned to useful purposes by an extensive infrastructure of local, regional, and national pipelines that now allow this clean, inexpensive, and easy-to-handle fuel to be used by business, industry, and residences around the country. Between 1945 and 1960, gas use increased significantly. It became the predominant fuel for residential heating and began to replace coal and oil as a boiler fuel for industry and electric utilities. Its low price and ease of use induced both residential and industrial users to become more dependent on it, and today natural gas supplies about one-quarter of domestic energy needs.

While oil and gas use was increasing at 3.5 percent per year (1950–1973), federal legislation was enacted to establish price ceilings for domestic natural gas and oil. The short-term effect of this activity was to decrease energy prices, which dropped 28

---

[*] These statistics were obtained from two sources: Executive Office of the President, *The National Energy Plan,* April 29, 1977 (Washington, D.C.: Government Printing Office); U.S. Department of Energy, *Synopsis of Energy Facts and Projections,* Annual Report to Congress, 1978, Document DOE/EIA-0173 (SYN).

percent in real terms from 1950 through 1970. The long-term effects, now clearly recognizable by all, were increased energy use, decreased domestic production, and increased importation of foreign oil. By 1947, the United States became a net importer of oil, but excess domestic production capacity exceeded imports. Prices for imported oil stayed low because of the discovery of large overseas oil reservoirs and the development of an efficient, economical system of international transportation. As U.S. consumption grew, U.S. production declined; and by the mid-1960s, we had become dependent on imports. Domestic production capacity could no longer match imports, setting the stage for the present crisis. Oil imports, which were negligible in the late 1940s, rose to 21 percent of total domestic consumption in 1955, 37 percent in 1974, and 48 percent in 1977. They decreased substantially to 38 percent in 1980.

In retrospect, we should have anticipated the problem. Even as our energy use increased and the shift from coal to oil and natural gas occurred, there were warning signs. Cheap imported oil combined with domestic price controls to lower energy costs. However, reduced prices and price controls shut off the more expensive domestic avenues of production, and for the first time, in the early 1960s, our rate of consumption exceeded our rate of discovery. The gap between discovery and production has increased every year since that time, with the sole exception of the Alaskan oil finds. Even a find as large as Alaska's only adds another 5 years to our supply. Domestic reserves, which were sufficient for a 14-year supply in 1940, have fallen to less than 10 years of domestic production, or about 5 years of present consumption. The rational economic answer to energy needs—turning to a large, cheap supply of imported oil—has turned out in the long run to be a poor political and economic choice.

The instability of the OPEC countries has been heightened by the influx of oil dollars, and tensions have been increased between the producing and nonproducing as well as the developed and nondeveloped nations. Of the developed nations, Japan in Asia and Italy in Europe are the most energy poor, and thus the most vulnerable. England is just tapping its own North Sea oil fields, while on the North American continent, Mexico and Canada have huge reserves of oil that provide them with self-sufficiency and some leverage in their relationships with the United States. Although U.S. oil supplies are dwindling, coal reserves can provide several hundred years of energy needs if social, technical, and political impediments to full exploitation of the reserves can be overcome.

ENERGY IMPACTS ON LOCAL GOVERNMENTS

Local governments consume huge amounts of energy. Vehicle fleets for police, fire, refuse collection, street and utility maintenance, and regulation, inspection, and service functions are visible daily reminders of how intimately fuel-consuming motor vehicles are tied into service delivery operations. Public buildings, office buildings, courts, jails, arenas, auditoriums, civic centers, maintenance facilities, and schools need heat and light, and increasingly, year-round climate control.

There are other less visible uses. Water treatment and supply plants, wastewater treatment plants, and even sanitary landfills are operations that require energy. Localities that supply water and treat wastewater are caught in a double squeeze. The EPA is mandating additional filtration for water supplies. This filtration imposes both a capital and an energy cost to push raw water through them for purification purposes. Federal regulations require all municipal wastewater treatment plants to provide secondary and, in some cases, tertiary treatment processes. These advanced processes are more energy intensive and in turn produce more waste that must be treated and transported.

Municipalities that enjoyed wholesale oil and gasoline rates several years ago are suffering. A price differential of 10 cents per gallon on fuel oil or gasoline when retail prices were 20 to 30 cents per gallon was substantial; a discount of 10 or 15 cents on prices in excess of a dollar a gallon does not provide the same margin of cost reduction. In addition, localities that were good bulk customers when fuel was plentiful have difficulty obtaining supplies at reasonable prices now that the supply–demand situation is much more tightly regulated.

Local governments are also losing preferential status in electricity rates as utilities are pressured to incorporate more equity, making sure that each customer pays an equal share of electricity costs. Areas with hydropower have cheap rates, but hydropower is approaching limits and any future growth must be served by fossil or nuclear plants. In areas where fossil plants supply power, preferential status, such as discounts, flat rates, and no demand charges, are rapidly giving way to commercial rate structures.

Specific trouble occurs when facilities that are high spot electricity users, such as convention halls, arenas, stadiums, and civic centers, are subject to demand charges. Demand charges are based on peak consumption during a nominal period, usually a half hour, and are charged by utilities to offset the cost of electric generation facilities that are not fully used but must be maintained to ensure that peak customer demand will always be met. These facilities are high users of air conditioning and lighting energy for relatively short periods of time, and demand charges are a heavy penalty for their use.

In addition, localities are buyers of materials and products that require energy in their production processes. Asphalt used for street paving is a particularly energy-sensitive material since it is a petroleum-based raw material that requires heat energy to put in place.

WHY WE SHOULD BE CONCERNED ABOUT ENERGY

The United States' energy use has grown about 3 percent annually since 1950. Despite some leveling due to recent dislocations and higher energy prices, our consumption is projected to increase at a 2 percent annual rate throughout the rest of the century. This has many implications. First, because of our dependence on imported oil, we are coupled to the instability of oil suppliers. Second, energy costs are a major contributor to the balance-of-trade deficit. Third, higher energy prices—a consequence of our relationship with emerging oil suppliers—has caused severe dislocations within our domestic economy. Fourth, there are near-term environmental and social consequences associated with the development of existing and proposed energy facilities as well as the development of new energy supplies. Fifth, there are global consequences associated with increased energy consumption.

The most pressing problem associated with our own energy shortage and our reliance on exporters, particularly the OPEC countries, is the political instability that the supply and demand situation has created. The dependence of the industrialized energy user nations on oil from less well developed supplier nations has concentrated enormous leverage in their hands. The solidification of these countries, through OPEC, has mobilized this leverage and translated it into political power and wealth. Although the most immediate consequences of OPEC action have been economic—higher prices, oil shortages, and competition among developed nations for oil—the most serious consequences arise from the concentration of wealth and power in the Middle East nations.

The major oil-producing nations of the Middle East are, by Western standards, sparsely populated and poorly developed economically, industrially, and socially. The sudden concentration of wealth has led to political instability in the area and to the possibility of more embargoes to using nations, political takeovers by nations or organizations not in sympathy with the United States, and ultimately, perhaps, a global war for control of resources. Ongoing instability in Iran, which was unable to make the transition to an industrial society, as Japan has over the last few decades, is a warning of future instability. Although decreased production of Iranian oil has not significantly affected the world situation, its impact will become greater in the next several years if world energy consumption resumes its upward trend, and an embargo by another oil producer will be enough to cause severe shortages.

Other politically unstable situations in this century have led to wars, and there is no reason to suppose that Middle East conflict will decrease in the near future. The most effective step we and other energy-consuming nations can take is to develop alternative sources of energy so that the OPEC countries' leverage and political importance can be diminished.

Our bill for imported oil exceeds $80 billion annually. This has decreased the value of the dollar abroad and caused instability in our own currency system. More importantly, this enormous transfer of wealth is a transfer of ownership and assets away from ourselves and into other hands. Although much of the capital is returning to the United States in the form of purchases—sophisticated weaponry, aircraft, and other manufactured products, as well as technical support services—it is also being invested in U.S. businesses and property. Although this capital investment benefits our economy, the money would benefit individuals in this country far more if it were retained in their hands rather than being spent for energy and returning as capital. The long-term effect is that we as a nation trade off tomorrow's returns on capital investment for today's oil cost.

There is some side benefit to the dollar devaluation—United States products are becoming more competitive with those of other nations. However, our companies have also become more attractive candidates for ownership by foreign investors because of deflated U.S. dollar values. A continued devaluation of the dollar can only have negative long-term consequences if we continue our shift from a manufacturing to a service-based country.

Within our domestic economy, the steep increase in imported oil prices and the steps we will have to take to remove subsidies and price ceilings on domestically produced gas and oil to stimulate new sources of fossil and nonfossil energy supplies will cause continuing dislocations. In the last several years, we have moved from cheap, declining energy prices to a situation of continuously increasing energy prices. Higher energy prices affect our entire economy—residential, commercial, industrial, and transportation sectors are all feeling its effects. Each individual can expect to pay more for energy to heat a home and run a car. In addition, the food we eat and the products we buy will have higher energy costs built in. Industries have seen oil prices quadruple and coal prices follow suit. Natural gas, which has been in short supply the last several years, is now indexed to the price of oil for industrial uses as a result of recent regulations. These costs are passed through as higher prices for manufactured goods. The impact on local governments and utilities is discussed in this chapter. They are facing the same turmoil. Overall higher energy prices mean higher prices for the economy in general, decreasing purchasing power, and spiraling inflation.

Increased oil production carries some environmental penalties, but oil has long been recognized as a vital part of our economy, and oil exploration has generally been well accepted. Some of the middle Atlantic states are fighting attempts at a federal level to lease offshore lands for commercial production because of fears of rapid development of shore-based production facilities and the influx of people, and because of the potential for oil spills from the producer wells and pipelines.

Full exploitation of western coal and shale reserves will cause severe restoration problems for the relatively fragile plains areas. There is also some question as to whether water supplies can support both energy production and population growth. Some western states are imposing significant severance taxes on their exported coal and are worried about boom-town impacts near energy supplies.

There is also substantial nationwide resistance to the development of energy production facilities. Nuclear plant construction has all but halted. Builders of refineries have had substantial opposition to construction along the Atlantic coast, and plants for generating electricity from fossil fuels are being opposed in the arid Southwest. Construction of pipelines to carry Alaska oil from the West Coast, where it cannot all be used, to the Midwest, where it can be used, has been so tangled in the federal, local, and state permitting process that the president of the United States intervened. Despite the intervention, the Sohio Company, which originally sponsored the venture, abandoned it. In all areas, the shift and growth of energy supply and production is encountering stiff resistance from those who are interested in restraining uncontrolled growth and protecting the quality of the environment.

The major long-term effect associated with the production of power by cumbustion of fossil fuels is the increase of carbon dioxide in the atmosphere. Although the Clean Air Act amendments of 1977 will produce a lowering of total particulates, hydrocarbons, and carbon monoxide in the air, sulfur oxide emissions will remain the same and nitrogen oxides as well as carbon dioxide are projected to increase through the year 2000 because of greater fossil fuel consumption.

Carbon dioxide (CO_2) is a major concern. Since 1950, CO_2 levels in the atmosphere have increased approximately 10 percent. A quarter of this increase has occurred since 1970. At present and projected rates of CO_2 emission, a two- or three-fold increase in atmospheric CO_2 levels could occur. It is estimated that average earth temperature will increase two to three degrees centigrade each time the atmospheric CO_2 level doubles. Although our knowledge of the increased CO_2 levels in the atmosphere is scant, the consequences are significant enough to warrant further study by the U.S. Department of Energy.

Energy is available to support the near- and far-term needs of the United States. However, the era of plentiful, cheap energy is over, and if we wish to ensure sufficient supplies throughout the end of this century and develop new sources for future centuries, it will be necessary to invest simultaneously in higher energy prices that will permit alternate fuels to be competitive; available but not completely developed technology that will produce alternate fuels; and fundamental and applied research to produce less demanding energy sources.

ENERGY ALTERNATIVES

This section will deal with major alternatives for meeting energy needs in the present and near future. The most realistic sources include the fossil fuels (oil, natural gas, and coal), atomic power, biomass (renewable crops that can be burned to produce energy), and solar and wind energy. Alternatives and adjuncts to these sources include synfuels, which are transformed energy, and energy conservation.

Oil and Natural Gas

For the last 20 years, domestic oil consumption has outpaced discoveries, even considering North Slope oil and outer continental shelf (OCS) exploration. However, some two-thirds of the oil has been left unpumped in existing domestic wells because price ceilings made it uneconomical to remove it from the ground. The gradual decontrol of existing oil prices over the next several years and the immediate decontrol of more expensive new oil from deep domestic wells should stimulate production. The higher prices— pegged to replacement rather than present production costs—will stimulate the use of secondary recovery techniques, such as heat injection, pressure injection, and chemical additives, to remove existing harder-to-get oil. Pricing should stimulate additional reserves in the short term. The best estimate for the long term is that most of the easy-to-reach oil has been discovered, and large new discoveries like North Slope oil will be less probable and more expensive. In the long run, we must limit oil use to chemical feedstock and high-priority transportation needs, substituting other energy wherever possible.

The domestic natural gas situation had until recently been distorted by federal regulation. Until the passage of the Natural Gas Policy Act of 1978, which will gradually decontrol gas through 1985, interstate gas was price regulated. Intrastate gas, which was not under federal purview, was allowed to float at free market prices. This had the effect of diverting 85 percent of the natural gas into non-price-controlled intrastate markets. As a result, demand for interstate gas exceeded supply. Interstate gas prices rose, but more slowly than intrastate prices and not enough to match supply with demand. Since the price mechanism did not work, federal and state authorities attempted to limit demand by curtailing service. The result was chaos. Natural gas reserves fell short of consumption in the 1960s, and gas production peaked in 1972. In the mid-1970s, however, gas production increased again, spurred by intrastate prices. Despite a sufficient aggregate supply, the gas was not available for the regulated interstate market, and industries have been cut off from gas every year since the mid-1970s.

Although the additions to natural gas supply may be a short-term trend, the presence of excess gas led to its decontrol. With prices for industrial users subject to immediate deregulation, there is now an excess of gas reserves and a better balance between supply and demand. New federal policy is to take advantage of what is believed to be a temporary gas surplus by diverting it to residential, commercial, and industrial uses to supplant 400,000 to 500,000 barrels a day of imported oil. At the same time, the Federal Fuel Use Act prohibits the use of natural gas in new utility and large industrial gas boilers. These boilers can effectively burn coal, and the act discourages large-scale use of gas where other fuels can easily be substituted.

The short-term outlook for gas is positive, and we are using it as a substitute for imported oils. The long-term outlook is clouded. It depends on drilling for new, harder-to-get gas and on the stimulating effect deregulation will have on existing supplies.

Coal

Coal is our most plentiful energy asset. Coal production, which peaked at just under 700 million tons annually in the period from 1915 to 1925, declined sharply through the 1930s. Its use rose again during World War II to just over 700 million tons a year and then declined gradually to a low of 400 million tons a year in 1960. Since then, it has climbed gradually back to the levels of the 1920s and 1940s. Coal, however, supplies only about 20 percent of our energy needs, and most of that is for electricity. In 1978, 78 percent of the coal used in the United States was burned by electric companies.*

Coal was supplanted by oil and gas because it is harder to handle. It is a solid and requires more complex mechanical handling equipment. Coal's chemical and thermal properties cannot be altered as those of oil and gas can. Different coals can be mixed to obtain desired properties—utilities often blend low- and high-sulfur coals to meet sulfur limitations, for instance—but this mechanical mixing is limited. Coals also leave an ash residue that must be collected and disposed of. Oils, by contrast, have very little ash content, and gas has almost none. Coal-burning

---

* These statistics were obtained from two sources: Executive Office of the President, *The National Energy Plan,* April 29, 1977 (Washington, D.C.: Government Printing Office); U.S. Department of Energy, *Synopsis of Energy Facts and Projections,* Annual Report to Congress, 1978, Document DOE/EIA-0173 (SYN).

facilities require handling equipment to remove the larger products from the combusion process; they also require equipment to remove particles from the host combusion gases that pass through the exhaust gas chimney or stack.

Some coals are relatively high in sulfur, and sophisticated, expensive (and not totally matured) techniques are required to remove the sulfur that oxidizes during the combustion process. All these factors influence coal's desirability as a fuel and weigh heavily in its ultimate price.

In addition to difficulties with its intrinsic properties, there are problems associated with its production and distribution that start at the mines themselves. Strip mining has been wrapped in controversy for years. The Appalachians are dotted with ruined areas where overburden has been removed with little concern for its consequences. Rainfall causes soil erosion and washes mine leachates into the streams, killing aquatic life. Deep mines do not cause quite the surface damage that strip mines cause, but the land above them may be prone to subsidence collapse caused by the removal of mined material.

Mining itself is a dirty, dangerous profession. Perhaps because of the work conditions, miners have developed a fatalism and stoicism all their own among industrial workers. Efforts to modernize labor activities and increase productivity are looked upon as management tricks. Union miners, particularly in the East, have engaged in long acrimonious strikes to better their livelihood, and even during times of overall labor peace, wildcat strikes of individual groups of miners are a common happening.

The nation's railroads, the major coal carriers, appear to be emerging from a long period of decline. However, during that long decline, rolling stock and the actual railroad beds themselves were allowed to deteriorate. The shortage of rolling stock can be overcome with a few years' additional rail car production. The reconstruction of track is a major capital investment that must be made to increase safety, speed, and load-bearing capacity, particularly in the crowded northeastern rail corridors. The railroad problem was severe enough to warrant federal takeover and reinvestment through two quasi-governmental corporations—Amtrak and Conrail.

Pressed by federal regulations, coal users are moving toward more sophisticated systems to remove contaminants from stack gases. Particle removal systems are fully matured, and most of the present development effort is associated with derivative systems that will remove sulfur as well as particulate. In addition, development is under way for processes that mix additives with the coal feedstock or that use additives in the combustion process itself.

The Department of Energy is funding gasification and liquification plants that have the potential for converting coal into a gas or an oil equivalent. These plants will produce easy-to-handle liquid or gaseous products that can displace domestic or imported gas and oil. This will allow existing transport and fuel-burning systems to switch supply sources with minimum impact. The rate at which these so-called synfuels reach production will be determined by the practicality of the plants being built and by the economies of synfuel production. Figures developed by the Department of Energy project that by 1990 synthetic fuels could compete with oil at crude oil prices of $20 to $45 per barrel, with gasoline at crude oil equivalent prices of $20 to $50 per barrel, and with natural gas at crude oil equivalent prices of $20 to $40 per barrel. Crude oil prices are in the upper-mid range ($30 to $40 per barrel) now, and if prices escalate at a rate similar to that between 1974 and 1981, the high range crude oil equivalent prices will be reached well before 1990.

Coal is our brightest prospect for solving our energy dilemma, but tripling or quadrupling our production to allow coal to become the major energy source over the next 10 to 20 years will require that we attack social, political, technical, and environmental problems simultaneously in the production, distribution, and combustion of coal.

Nuclear Power

In the longer term, past the year 2000, present-day nuclear reactors must be viewed as a stopgap. The fuel they use is scarce and expensive to produce, and there are problems of disposing of it when it is spent. The newer breeder reactors will produce more fuel than they use, and they have the potential, along with nuclear fusion, to be a long-term answer to our energy needs.

In the interim, however, we must concentrate on solving the social and technical problems associated with current technology, since we would be unable to meet the present needs for electrical energy if nuclear plants, which account for 13 percent of domestic power-generating capacity, were removed from service.

Social hysteria, amplified by the reactor failure at Three Mile Island in Pennsylvania, must be calmed by a combination of better regulation and surveillance by the Nuclear Regulatory Commission or its descendants, and more attention must be paid to safe design and adequate operator training. The Department of Energy has instituted a national program to provide for the safe disposal of nuclear waste from power plants (as well as waste from military, medical, and foreign uses). This program will be spread over many years and includes provisions for the development of two or three sites in the continental United States. The pace of the program will be dictated by institutional as well as social issues, since nuclear waste depositories are not considered desirable by any locality.

The overwhelming regulatory process associated with siting and building nuclear power plants has undoubtedly contributed to the operational difficulties nuclear plants are experiencing. Any process that drags on for 10 to 12 years is bound to shift focus from basic design and construction to a continuous leaping of bureaucratic hurdles, siphoning off inspiration and dedication in the process.

Lower than projected demand for electrical power, coupled with dramatic increases in plant construction cost, along with an extensive regulatory process and citizen fears, has brought nuclear plant construction to a standstill. If nuclear power is to assume a larger role in the overall energy strategy, then we must press hard to remove the existing barriers to plant construction and operation. Interestingly enough, the viability of nuclear power rests on such issues as how we govern and regulate ourselves and on public perceptions of the dangers associated with atomic power. Technical considerations, although substantial, will have little to do with the future of nuclear power over the next 20 years.

Renewable Resources —Solar

Despite its present social popularity, solar energy is not yet an economically viable alternative energy source. The use of direct solar energy to produce heat energy is more mature than the conversion of solar energy directly to electrical energy, even though the latter has more inherent flexibility and potential.

Existing solar thermal conversion systems, although simple in concept, require much material for panel construction and are thus high-cost items. Although solar energy becomes more attractive as fossil fuel prices rise, present-day solar collector materials—glass, aluminum, copper, and insulating materials— are themselves energy intensive to produce, and collectors are caught in the price spiral. Two of the major conditions for successful integration of solar thermal energy into our system—large-scale production, and materials and cost breakthroughs that can only result from extensive research—cannot take place overnight and cannot be produced by federal regulations. Major breakthroughs in low-cost materials are required, as are simpler, more efficient ways of producing the higher fluid temperatures necessary to make solar-powered air conditioners practical.

In the long run, solar electrical conversion will provide more direct benefit than thermal energy. It is a more flexible energy form, since it can easily be tied into the electrical distribution system of a building and be converted to power or thermal energy. The widespread use of solar electrical energy will cause, if anything, a more significant impact on power-generating load profiles of electric utilities than will the use of solar thermal energy. Utilities must be prepared to generate replacement power during nighttime and cloudy or overcast days. Although overall solar energy will decrease net fuel consumption, it may not decrease energy cost because of its stop-and-go impact on power-generating utilities.

Biomass

Biomass energy can be classified as that energy developed from materials that are the result of a growth process as opposed to the very slow decomposition process that produces coal, oil, and natural gas. Energy is derived from the combustion of materials that have been produced by the absorption of the sun's energy. Biomass materials are renewable, since they can be grown, harvested, and replanted in a continuous cycle. The cycle varies—some hardwoods take a half century to grow to full maturity; corn and sugar are reproduced annually.

Wood has been supplanted by other fuels only in the last hundred-odd years, and it has served as a renewable fuel from the time that civilization began. It is not quite so amenable to large-scale production and distribution as the fossil fuels, and also it has substantial value as a natural resource in other areas. It is the most widely used structural material in the construction of homes, buildings and furnishings, packaging, and other areas where its ease of working, structural strength, and appearance are needed. Its fibers are used for the production of paper and other fabrics. In addition, wood serves as a source of chemical feedstock for lubricants, food additives, and fragrances, to name but a few products.

As other fuels increase in price, wood again becomes attractive, despite difficulties in collecting and handling. The major impediment to its widespread use, however, is its versatility. Excessive consumption as a fuel will cause price and supply stress in other industries that use it as feedstock, and substantial conversion to wood as a fuel will cause dislocation in many other areas of our economy.

Corn, much in vogue as a source of ethanol, which is used in gasohol (a 90 percent gasoline, 10 percent ethyl alcohol fuel), is similar to wood in that it is a crop. It suffers from the disadvantage that the fermentation of corn to alcohol and the subsequent distillation of the alcohol to the higher purity needed for gasohol requires about as much energy as the alcohol itself produces. At best, the production of ethanol using nonpetroleum fuels can result in a shift away from imported oil. At worst, its use can cause a net energy loss that must be subsidized to make gasohol an attractive fuel in the marketplace.

More research is needed to evaluate crops that can be more intensively cultivated, that can produce more energy, or that are subject to less demand for use as foods. With annual sugar consumption decreasing in developed nations, sugarcane may make an attractive biomass crop. More research is also required into multifaceted crop uses, such as the use of residual materials from the alcohol production process as animal feeds. Biomass is a largely untapped, attractive source of energy that has the benefit of being renewable. It also makes a good fit with land availability and domestic crop-growing practices.

Synfuels

Gas and oil dominate as fuels because they are convenient to transport and burn; they can be carried by truck, rail, or boat or can be pumped in a pipeline; and they burn easily, with a minimum of stack pollutants and of residue. Our energy distribution and use network has been built around these features, and the conversion of coal, wood, and other solid fuels to liquid or gaseous forms, and the creation of synfuels, would make it possible to substitute synfuels for oil and natural gas with a minimum of disruption to our existing systems. The conversion of coal to liquid fuel is not a new process—Germany did it in World War II when the allies cut off natural oil sources, and South Africa has had production plants in operation for several years. However, the production of oil and gas from coal has only recently become a reasonable economic alternative, and the U.S. Department of Energy has large-scale technology demonstration programs under way to reduce synfuel capital and operating costs and to solve the health and environmental problems associated with synfuel production.

CONSERVATION

During the mid-1970s, there was much confusion in the minds of the public about our energy situation. The media, elected officials, government agencies, oil companies, and various prophets and saviors inundated us with information that was confusing and often conflicting. The continued drastic rise in OPEC oil prices and some constructive federal policies that will gradually remove domestic oil and gas price ceilings have eliminated much of the confusion by reducing the situation to basic economic terms: Oil and gas have become very expensive, and the expense warrants each individual's attention.

The first indication that everyone is taking costs seriously was the sudden shift to more energy-efficient cars when gasoline prices passed above $1 per gallon in 1979. Although the domestic demand for small cars had been growing, it took a step jump in 1979, and in view of the three to four years the auto industry needs to convert its assembly lines, the change must be judged to be sudden. A good subjective indicator of the shift is that small-size American automobiles are commanding better prices than their immediate-size counterparts. Gasoline cost is now a major consideration for the purchaser.

Local governments have responded by purchasing smaller vehicles. Compact patrol cars previously eschewed by police departments are now common, as are imported pickup trucks.

The next major conservation shift will be triggered in the residential and commercial real estate markets, as energy costs for natural gas—the cheapest and most convenient (and therefore most widespread) heating fuel—double during the 1980s because of gradual deregulation. Heating and air conditioning costs, which were once very low, are now approaching annual capital costs. As owners feel the pinch, the variable costs represented by heating and cooling will get more attention, and such energy conservation methods as insulation; more efficient heating, ventilating, and air conditioning systems; and better controls will make good economic sense.

Energy conservation efforts in public buildings will continue to move slowly. Schools and hospitals, particularly, are large energy consumers that are shielded by a very conservative set of regulations to protect the public. Hospitals are among the most energy-intensive buildings, and care must be taken not to trade patient safety for energy dollars. However, there is room for more efficient heating and air conditioning systems with no sacrifice in overall health effects, and there is a need to overhaul attitudes as well as codes.

Schools present a slightly different set of conditions. Schools generally account for 50 percent of a local government budget—a strong testimonial to the importance of education in our culture. This economic investment is backed up by a strong attitudinal investment that shows up in the conservative codes governing school buildings and buses. There is much room for improvement and innovation in school building design and operation and in school bus design, but reluctance to focus on activities that are not curriculum related and long-standing conservatism rooted in health and safety considerations make progress in energy conservation slow.

In recognition of the difficulties associated with changes to schools and hospitals, the federal government has instituted an energy grant program for these institutions. There is also some sentiment to shift away from conventional school bus design in favor of cheaper, more efficient vehicles.

SUMMARY

We are in a state of transition in terms of energy supply and use, and it will be difficult to predict our exact direction. Massive federal studies in the early 1970s, triggered by the Arab oil embargo, were unsuccessful at predicting both the costs of imported oil and the types of dislocations OPEC pricing policies would cause.

There are some encouraging signs, however. Total oil consumption is down and worldwide oil imports are off, causing the first large-scale drop in OPEC-set prices in 10 years. Although part of the drop may be attributed to the present recession, much of it is due to conservation activities in all sectors of the world economy and a shift to alternative fuels. After years of vacillating, federal decision makers have decided to remove price ceilings from domestically produced oil and gas. The result is that the market will drive consumers to conserve and that alternate fuels will become more competitive. The first sign of pricing impacts has already occurred: There has been a strong shift to smaller automobiles. Industries are again considering more extensive use of solid fuels, cogeneration, and waste fuels as alternatives to oil and gas.

There are also some long-range concerns about alternatives. Nuclear power is clouded with social and safety issues. Conversion to coal carries environmental penalties, and our transportation system needs modernization if it is to assist in the conversion to solid fuels. And, of course, the transition from cheap to expensive fuel will create substantial stress for individuals and institutions. But the implications of continuing down our present path—political blackmail, transfer of wealth, and the possibility of global war—make it necessary for us to make the investment in energy self-sufficiency and stability.

Local governments are particularly vulnerable to the energy price squeeze. They are large users both of energy for space heating and cooling and transportation and of petroleum-based materials, such as asphalt. The price inflation in energy manifests itself in higher materials costs, and in higher salaries as employees fight to stay even with costs. Tax revenues are lagging, and localities find themselves capital poor at a time when capital investment is required to decrease energy costs. And even when money is invested in energy-saving systems—modified heating and air conditioning systems, automated building controls and fuel dispensing systems, and the like—energy savings often fall short of projections. Localities will suffer as greatly as any institution from our energy situation in the short term.

In the longer term, prospects are brighter. New and rebuilt buildings and fleets will be more energy efficient. There will be opportunities to use automation and to reduce personnel and transportation costs. A swing to mass transit to decrease personal transportation costs will favor urban areas. And energy from solid waste will provide opportunities for localities and their industries to benefit from waste material. The path will require more careful assessment of our present and future options in order to trade technology for energy costs on our way to energy independence.

CHAPTER 10

The Revolution
in Electronics

Energy and environment will be significant driving forces in the 1980s because they will establish limits on what we can and cannot do. In a sense they are negative forces—energy prices forcing costs up and intensifying the search for alternatives, and environmental laws limiting or eliminating practices that have been found harmful to people and the air, water, and land they use. Electronics, by contrast, will be a positive force, enabling us to calculate, communicate, command, and control better and less expensively than ever before.

The electronics revolution has been marked by the convergence of several technologies that evolved separately but are now combining to form one integrated scheme of information processing and communications. Computers provide the basis for information processing; radio and television are the communications media. This chapter will trace the development of communications and information processing technologies to their point of merger. It will then focus on the importance of communications and information to local governments and on the opportunities and problems that will be presented as these fast-moving technologies merge and evolve.

COMMUNICATIONS

The communications industry was originated in 1832 when Samuel Morse invented the telegraph, a means of conveying signals over wires in a binary fashion by switching voltage on and off. Morse code translated this binary system into letters and words, and in 1844 Morse received a U.S. government grant to install his system between Baltimore, Maryland, and Washington, D.C.—a distance of 37 miles. The first transatlantic telegraph cable was installed between the United States and Great Britain in 1866, making transoceanic communication possible.

Some 10 years later, Alexander Graham Bell invented and patented the telephone, a device for carrying voice signals over wire. Bell's first commercial unit was installed in a Boston bank, and the Bell Company, formalized in 1877, was on its way to spectacular growth.

During the same period, the first storage device, the phonograph, was invented by Thomas Edison. In addition, several other components essential to the development of the radio—the carbon microphone, the loudspeaker, and the aerial (antenna)—were developed. Edison built on the work of others, including Bell, to develop the carbon microphone in 1877, a device still in use today. In the same year, E. W. Siemens in Germany developed the mechanism for the loudspeaker, which was ultimately perfected in 1925. In 1887, Heinrich Hertz used the first antennas to demonstrate radio propagation. In 1896, Marconi patented wireless telegraphy, a coupling of radio propagation with telegraphy techniques, and formed the wireless telegraph company that transmitted transatlantic signals in 1901.

At the beginning of the twentieth century (1904), the first electron tube, the diode, was developed as a result of Sir John Fleming's work with Edison. This was followed two years later by Lee deForest's invention of the triode—a device that allowed small electrical signals to control large electrical signals, producing the first practical electronic amplifier. At the same time, the first successful radio broadcast of music and speech was accomplished in Massachusetts, using a microphone in the antenna circuit of a 1,000 watt, 50,000 cycle per second generator developed by the

General Electric Company. Demonstrations followed in New York City in 1907 and in Paris in 1908, ushering in the radio broadcast era.

In the 1920s, theories for television, the broadcast and reception of visual (video) information, were worked out, leading to the development of the iconoscope, a video camera, in 1923 and of commercial television in the 1930s.

In the same period, the first commercial shortwave broadcasts were made at a frequency of 11 megahertz (11 million cycles per second) from the Netherlands to a Dutch ship off Indonesia, and long-wave transmission facilities became obsolete. In 1933, E. H. Armstrong developed frequency modulation (FM), a noise-free alternative to the then standard amplitude modulation (AM) of radio waves. The intervening period of the 1930s and 1940s was a time of rapid development and commercialization of radio technology for all sectors of our economy: commercial broadcast, military uses, and two-way radio for private and government use.

A major technology breakthrough occurred in 1948 when Bell Laboratories developed the transistor, a solid-state device that performed the function of the vacuum tube. Vacuum tubes use thermionic emissions—the controlled release of electrons from a heated emitter—as their principle of operation. High heat and power are required to cause thermionic emissions, and electron tubes are large, expensive, and, by today's standards, short-lived devices. The transistor works on the principle of using electrical currents to cause semiconducting materials to conduct electricity and accomplishes the same amplification with orders of magnitude decreases in power and size. The resultant decrease in size, power consumption, and cost coupled with increased reliability and capability paved the way for the revolution in solid-state electronics.

The first transistor radio set was introduced in 1954, and in 1959 the first microcircuit—a complete circuit contained in the body of a semiconductor—was patented. This was followed in 1960 by the development of linear integrated circuits, which largely substituted transistors and diodes for resistors and capacitors so that self-contained amplifier circuits with a minimum of external components could be employed in electrical circuits. This technology led to the development of the printed circuit board, where wires are replaced with flat metallic conductors etched onto an insulating material to form a prewired board to which integrated circuits and other components can be attached.

As circuits and construction techniques advanced during this period, so did the radio spectrum. Pushed by channel requirements and aided by increased knowledge, designs moved frequencies successively from high frequency to very high frequency (100 megahertz) to ultrahigh frequency (UHF, or 500 megahertz) and into microwave frequencies (800 megahertz and above).

At the same time, telephone communication was starting to be influenced by the computer. In 1960, the Bell Telephone Company installed the first electronic switch, a mechanism that uses computers and solid-state devices to create a flexible electronic switching path, supplanting fixed wired switching systems. The advent of an increased frequency spectrum, combined with the laser as a coherent light source, led to the modulation of light as a transmission medium, and in 1966, the theoretical groundwork that has led to the fiber optic cable was developed.

From a practical point of view, radio and telephone have merged as a mixed medium in many areas of communications. Telephone companies take advantage of the signal-carrying capability of microwave systems to relay messages over long distances. Radio communication takes advantage of the efficiency and economy of telephone systems to carry audio signals from multiple origination points to a transmitter where the traffic density or communications diversity justifies the use of the limited frequency space available.

COMPUTERS

The first computer, a mechanical calculator, was developed in 1833 by Charles Babbage, the father of industrial management. This so-called analytical engine was a series of gears, drums, and levers capable of performing the functions of memory, control, arithmetic, and input and output. This mechanical device, far ahead of its time, set the stage for twentieth-century electronic machines.

In 1889, H. Hollerith developed a punched card tabulating machine to gather statistics for the 1890 United States census. Individual census data were represented by holes punched in cards, and these holes were counted by pins attached to counters as the cards were passed through the tabulating machines to produce totals for each piece of census data gathered. The tabulating system was a major advancement in that it reduced the census tallying process from months to days.

The next major step was the development in 1931 of the differential analyzer, an analog computer for solving differential equations. The analog computer suffers from difficulty in signal reproduction and time and frequency interactions, but it is still used for specialized control and simulation to this day.

The first digital computer, the Mark I, was developed by IBM between 1939 and 1944. The Mark I was an electromechanical calculator, 51 feet long and 8 feet high. It could add, subtract, divide, and refer to tables of previously calculated results. The machine was controlled by an automatic sequencing mechanism. Its input was by punched card or switch positions, and its output was by punched card or electric typewriter.

In 1942, The Moore School of Electrical Engineering at the University of Pennsylvania started development of ENIAC (Electronic Numerator, Integrator, and Computer) to perform computations of military gun firing and ballistic tables. The unit was completed for the Aberdeen Proving Ground of the U.S. Army Ordnance Corps in 1946 and was used to perform integration of differential equations. The unit occupied a space of 30 by 50 feet and contained 18,000 vacuum tubes in place of relays.

This unit was followed by EDVAC (Electronic Discrete Variable Automatic Computer), which was built by the Moore School between 1947 and 1950 for the U.S. Army Ballistic Research Laboratory at Aberdeen Proving Ground. It contained 500 vacuum tubes and 12,000 semiconductor diodes and used a binary rather than decimal number system in its calculations.

The UNIVAC (Universal Automatic Computer), which was built between 1947 and 1951, was a direct descendant of ENIAC and EDVAC. The unit had 5,000 tubes and several times as many diodes and featured internal memory provided by mercury delay lines. Input and output were through a metallic-based magnetic tape. The first unit was built for the U.S. Bureau of the Census, and ultimately, 48 machines were built.

The UNIVAC was built by a spin-off company formed by two researchers at the University of Pennsylvania. That company merged with the Remington Rand Business Machines Company, which subsequently merged with the Sperry Corporation. The original company is now the UNIVAC Division of the Sperry Rand Corporation.

In 1954, IBM introduced the IBM 650, an intermediate-size vacuum tube computer, which became an industry workhorse during the late 1950s. Over a thousand of these units, which had magnetic drum storage, were placed in service.

In 1952, IBM and MIT's Lincoln Laboratories, in a cooperative effort, began development of a military communications–oriented computer, SAGE, which accepted radar data over telephone lines, processed the data, displayed information for operator decisions, and guided interception weapons for the Air Force. The first production unit was delivered in 1956.

In the mid-1950s, IBM developed the 700 series of processors. The 704, delivered in 1956, featured high-speed magnetic core memory. It was followed in 1958 by the 709, which had simultaneous read, write, and compute capacity. The 7090, first delivered in 1959, was a transistorized system comparable to the 709 but twice as fast.

The totally solid-state UNIVAC System 80/90 was introduced in 1960. This medium-size processor used solid-state devices for memory as well as computations.

In 1961, Digital Equipment Corporation introduced the minicomputer. Minicomputers were less powerful than the then existing central processors. They used smaller digital words, and they had less memory, but their low price made them attractive, and they found many applications. Over the last 20 years, the use of integrated circuits has reduced their cost by an order of magnitude and produced substantial increases in performance, blurring the demarcation between minicomputers and central mainframe processors. They now come equipped with a long list of sophisticated support equipment, and they can be linked together to form a distributed processing network that provides the aggregate processing power and speed of a single large mainframe processor.

The minicomputer was followed in 1977 by the microcomputer, which was developed by the Intel Corporation. This unit had less capacity than the minicomputer, and less flexibility, but it was smaller and lower in price and has found use in a wide range of applications, such as data communications, process control instrumentation, automotive vehicles, and consumer products, where simple repetitive decision-making capability is required. In 1976, Intel developed a one-board microprocessor, which contained the processor, memory, and input-output capacity on a single printed circuit card.

In parallel with microprocessor development, low-cost peripheral devices have evolved. In 1970, almost as an afterthought, IBM introduced the floppy disk, a low-cost data storage device, to perform diagnostics and some memory disk drive control on its System 370. In 1973, the floppy disk was introduced as part of a smaller data entry system and became an overnight standard. At the same time, development was proceeding on the Random Access Memory, a low-cost device for real-time solid-state storage that is the heart of the microprocessor.

As the processors became smaller, cheaper, and easier to use, they made their way out of the computer specialist's domain and into the user's domain. Terminals with limited intelligence and memory were placed at the point of data entry and manipulation, and a network providing access to, and transactions with, more powerful machines having large storage and high processing speeds evolved. These networks are connected largely through the private and public telephone network, once again integrating the computer and communications.

LOCAL GOVERNMENT NEEDS

Local governments as large-scale public service delivery corporations have a vital need for communications. Good contact between citizens and a locality are vital for responsive delivery of services and for citizen confidence. Telephones are used to provide a closed service link for calls for service and the resolution of problems. Two-way radios are used to dispatch field crews to handle complaints and respond to requests, and to mobilize personnel in the event of such emergencies as fires, flooding, traffic accidents, water and gas main leaks, and electricity disruptions. Broadcast media are used to announce public hearings and meetings, provide information about local government services, and explain local issues.

Internally, telephone communications and two-way radio are used to provide interdepartmental coordination and service dispatch. A two-way radio is now considered essential for efficient dispatch of emergency services (police, fire, and medical) and almost as essential for providing rapid response to problems in the utility services that local governments provide.

Information processing provides the internal operations fabric for local governments. Computerized financial management systems, which provide for record keeping in such areas as tax and fee collections, budgeting, purchasing, payroll, and other cash disbursements, are all but universal in large cities and counties and are finding their way at various levels of sophistication into all units of local government. Sophisticated information processing systems can provide closer control over receipts and expenditures, faster access to information related to those transactions, and better, faster information necessary for day-to-day conduct of the business affairs of a local government.

The early promises of vast cost savings from automation of local government systems have not materialized, however, particularly where complex combinations of people and procedures are involved, and many governments find themselves saddled with large equipment investments, substantial technical staffs, and seemingly little return on investment. The combination of fast-paced technology change and far-reaching impact on people and systems makes data processing the single most difficult area of activity for local government decision makers and managers. And the increasing merger of communications and data processing presents larger opportunities and problems that must be approached with more care than ever before. The potential benefits are large; the investment in people and facilities will be substantial as well.

TELEPHONE COMMUNICATIONS

Telephone service, which used to be a simple, almost noncontractual agreement between a locality and the telephone utility, has become a complex and in some cases controversial issue. Decisions in federal courts have now made it possible for an individual or corporation to own or lease as well as rent telephones and telephone switching systems. Telephone operating companies are being subjected to the same public scrutiny that power companies are receiving, as these franchised monopoly services increase in price in response to inflation and rising energy, materials, and labor costs. The telephone operating companies are now faced with competition, in terms of both providing telephones and switching systems and providing long-distance message services.

In response to these pressures, the telephone companies, long restricted from information services and data processing operations, are looking to those operations as natural extensions of their communications functions. In 1982, AT&T reached an agreement in principle with the federal governing agency, the Federal Communications Commission (FCC), for additional powers in these areas in exchange for a divestiture of local operating companies. Several corporate data processing giants, in turn, are looking for an expansion of their capabilities into the previously protected land communications areas. These activities, symptomatic of the merging of communications and information processing technologies, portend large-scale changes in the policy and regulatory arena that will have a profound effect on how communications and information processing technologies emerge over the next two decades.

In telephone services, local governments are faced with technology choices and supplier choices. All telephones need switching systems to allow interconnection with other telephones. These switching systems range from small push-button telephones (key systems) to large installations at telephone company central offices, where portions of a larger switching machine are used to provide interoffice, local, and long-distance calling capacity for an office, building, or organization.

For many years, the switching functions have been accomplished by crossbar switches, which use a fixed matrix of connections to route a signal from one telephone to another. In a simplified illustration, the crossbar switch can be viewed as a series of horizontal bars, each representing a telephone, capable of being connected to any one of a series of vertical bars, each representing a telephone trunk line. Electrical relays cause the horizontal and vertical bars to touch, making the connection between telephone and trunk. At the other end of the call, the trunk is connected to the called telephone and the call connection is completed. This technology is mature, providing high reliability at low cost. Its principal liabilities are lack of flexibility and large size.

The processor-controlled switch, the latest in telephone technology, uses solid-state electronics and computer control to decrease size and increase flexibility and capacity. The computer performs the routing, using a random path in place of the fixed path used in crossbar switches. The same processor is also used to perform time division multiplexing to save space and complexity. In time division multiplexing, voice signals from several sources are sampled, interleaved in sequence, and transmitted over a single pair of wires. Since only 5 percent of a voice signal is required to convey all the information needed for transmission, 16 or more signals can be multiplexed over a single set of wires.

The computer can also perform such functions as selecting the cheapest routing for long-distance calls, forwarding calls to another telephone in accordance with preprogrammed rules or instructions provided by individual telephone users, automatically redialing calls when busy lines become open, and a host of other features. Because of the processor control, telephone identities (telephone numbers) are assigned by the processor instead of by the wiring path. This makes it possible to reassign telephone numbers by reprogramming, rather than rewiring, and to prewire a building so that any combination of telephones can be added or eliminated through a combination of programming and minor wiring changes.

All the features outlined for electronic processor-controlled switches can increase communications capability and decrease cost both for day-to-day operations and for the ever present changing of offices and functions. However, electronic switching systems are high in initial cost, and prices vary considerably depending on the supplier. Thus, substantial matching of departmental requirements with system capability is required to realize the savings the electronic switches can offer.

If a local government decides to locate the telephone switching system in its own building, the choices become even more complex, since an ownership option becomes available. The Supreme Court declared in the *Carterfone* decision that telephones and switching systems from other franchised telephone operating companies could be connected to the operating company's system. This opened the way for so-called interconnect companies, private companies that sell or lease telephones and switching systems and provide maintenance service to their customers.

The interconnect companies offer the advantage of outright ownership or lease in place of rental. Although ownership does not give local government the tax advantages available to private sector owners, interconnect equipment can provide substantial price savings and capability advantages over telephone operating company equipment. In response to the interconnect challenge, telephone utilities are now offering long-term contract lease options and opportunities to exchange capital payments for lower operating charges as alternatives to traditional straight rental terms.

The choice between interconnect and operating company is complex. The decision to purchase from an independent supplier implies a long-term contract for maintenance and replacement parts, and it means a careful evaluation of the company's stability, future prospects, technical skills, maintenance crew size, and financial position. In addition, the question of equipment obsolescence is important when ownership is contemplated. Localities must also recognize that telephone operating companies are regulated utilities with a function parallel to their own, and some consideration must be made to ensure the continued health of this essential service organization.

Selecting telephone service for a local government has become a complex process, involving substantial capital investment, increasing operating cost, difficult choices in a fast-moving technology, and ever wider choices in lease, ownership, and rental as well as a choice between utility or private sector services. All these elements dictate a careful assessment, and Chapter 11 outlines the decision-making process that went into the selection of one county's telephone system.

RADIO SYSTEMS

Two-way radios have become the lifeblood of local government dispatch operations, particularly for emergency services. For years, radio dispatch of police cars in response to calls for service has been a routine operation. Fire services use radio communications for dispatch, en route coordination, and fire scene combat operations. Medical services use radio communications to telemeter patient medical data and coordinate emergency treatment with hospitals. Public works and public utilities operations use radios to dispatch field crews in response to minor problems, such as road patching and water and sewer main leaks, and to ensure coordinated response to major problems, such as flooding, natural disasters, and major utilities problems. Building and facility maintenance operations, field inspectors and meter readers, and other operations use radio communications to coordinate work center and field operations.

The single most important advance in two-way radio communications has been the development of the hand-held portable radio. Early field radios, incorporating vacuum tube technology, had to be vehicle mounted because of their size, weight, and power demands. Solid-state components, particularly integrated circuits, have reduced size, weight, and power demands to the point where a radio weighing less than one pound and fitting in one hand is capable of operating reliably for a whole work shift with no decrease in output power and communications capability. This allows the radio communications function to follow the user instead of the vehicle, providing on-the-spot communications to a police officer responding to a call for service or a fire fighter involved in fire combat.

Portable radios are limited in power output. Five watts is a practical limit, compared to 100 to 250 or more watts of output power from a base station. In addition, channel crowding has moved public safety communications into higher frequencies, where signal propagation and penetration are lower. Because of these limitations, radio systems have moved toward a satellite receiver concept, spotting remote receiver sites around a locality as required to ensure a maximum probability of receiving a transmission from a portable transmitter no matter where it is carried. The satellite receivers in turn are connected to the base station by telephone lines or additional radio frequencies so that the transmission can be received by the dispatcher and rebroadcast over the radio network. This communications system provides more reliable radio transmission and reception and instant response, but requires two frequencies for each communications channel and a large number of satellite receivers and antennas. It is a step jump in equipment, cost, and complexity and requires careful design to ensure good use from the dollars and equipment invested.

In the radio dispatch center, the computer has made drastic changes in some emergency service operations. The dispatch console is the focal point for operations, and it is at this point that telephones, two-way radios, and computers merge. The telephone provides the inward path for request for services from the public. It also accesses nongovernment support services and provides an alternate communications path. It often provides primary notice to station-located support operations, such as fire trucks and ambulances, that services are required. Two-way radio communication provides the dispatch function previously referred to and also serves as the coordination medium for interservice activities. The computer has found use as an information depository, a resource organizer, a status and record keeper, and a communicator between information networks.

A computer-aided dispatch system uses a cathode-ray tube (CRT) display and keyboard to extend the communications dispatcher's capability. In its most common form, it provides a status display of available emergency service personnel and vehicles, and a running history of the status of calls for service. It is also widely used to access data banks, such as the state vehicle licensing system and the National Criminal Information Center (NCIC), permitting access to a nationwide set of information files that provide information about people, vehicles, and other property. The status and information access capabilities provide the field response units with fast, accurate information necessary for the discharge of their tasks and assist in allocating resources.

Computer-aided dispatch systems can also be augmented by computer-based geographic data base systems that pinpoint addresses and can carry supplemental information about occupants or building contents that can be useful to police, fire, or medical services. The dispatch information entered into the computer can also provide the basis for a historical file that originates with the person, incident, or property involved in the service call, and, of course, the dispatch data can provide information for subsequent hourly, daily, monthly, or annual detailed reports and analysis.

This merging of technologies provides a substantial opportunity for efficient response to citizen needs and maximum use of service personnel. It also requires a substantial outlay of capital and operating funds. Integrated communications consoles are expensive items of custom-tailored hardware. Dedicated or shared computers must be fairly large to accommodate on-line transactions. Custom-tailored computer programs require substantial development, and data files require continuous input and update if they are to be of value. A communications center of this type has a heavy concentration of radio communications equipment, telephone switch gear, computers, and such peripherals as CRTs and printers. Substantial employee or contractor maintenance skills and a high level of competence among communications personnel are needed to ensure optimum system use.

Given the unique needs of each locality and the range of equipment and organizational options, there is no standard system that can be purchased off the shelf. Considerable foresight is required to develop a system that is flexible, cost-effective, and capable of being integrated with other local government systems and operations.

DATA PROCESSING

The digital computer has become an indispensable but expensive part of local government operations. The magnitude of financial activities—levying and collecting taxes; preparing and administering budgets for departments, agencies, and utility enterprises; receiving and tracking income from diverse sources; and accounting for purchasing and other expenditures—has mandated financial management systems, particularly in larger local government units. Payroll systems are all but universal, and utility billing systems have become commonplace, as have systems for automating licensing fees and other kinds of fees and permits.

At the outset, large multipurpose computers were the rule. Systems for supporting local government operations were scarce, and large-scale start-up activities and staffs were and still are required to support and expand the systems. A large investment in equipment and staff is required to support extensive data processing activities, and the promised decreases in costs have not materialized.

It is now commonly recognized that automated systems save search and access time and provide more accurate data and better record keeping, but they cost more than manual systems. Centralized, large-scale data processing systems still suffer from excessive rigidity. They require information input in a very specific format, and changing output reports or accessing data in a different way can require expensive and time-consuming programming changes, even if much of the desired information is contained in the information collected. Comprehensive data base systems, which cut across departmental boundaries, gathering and outputting shared information, are still very difficult to implement.

The rigidity of input and output requirements, privacy issues, the difficulty in deciding which information is valid and who has a right to revise it, and the massive changes automation can make to systems that have evolved with time combine to make data base information systems difficult to construct. Technology limitations have to be accommodated, and manual systems have to be completely characterized so that the automated system provides all that the previous system provided while at the same time streamlining operations. Any system that involves people must make allowances for their responses to it and for a natural tendency to resist change. For these reasons, large-scale integrated data base systems will continue to present a challenge to data processing professionals and information system users alike.

In the early 1970s, the minicomputer appeared. This device is smaller and simpler than central or mainframe computers. It has less memory capacity, uses less complex programming languages, and has simpler operating systems and less peripheral equipment. It also features lower cost (less than $20,000 when it first appeared in 1974). These characteristics make minicomputers ideal for dedicated tasks within a department or section. They are easier and more flexible to operate and, when coupled with a CRT readout for information display and a keyboard for data entry, they form the nucleus of a new approach—distributed data processing. Distributed processing is more timely and flexible than centralized data processing, which requires discrete steps of data input, information processing, and data output. Distributed processing also competes strongly with remote terminals connected to a central computer to provide on-line processing.

Minicomputers can now be purchased with software for dedicated, single-purpose systems that are suited to local government departmental needs. Dedicated programs feature preprinted, fill-in-the-blanks formats and instructions that lead the user step by step in a question-and-answer format through an operation as required for ease of use.

The microprocessor is yet another evolution in data processing. This microcomputer has more limited memory, and it usually contains fixed or simply programmed logic. Programming language is more elemental, and the operations are more specific and less flexible. However, complete microprocessors are now contained on a single circuit board, and when combined with limited peripherals, they can be powerful, low-cost devices, particularly for performing repetitive operations that require limited decision-making power. The microprocessor can easily perform many of the information processing functions that day-to-day local government activities require. Microprocessor-based terminals can perform as stand-alone processors or can function as intelligent input-output terminals for minicomputers or central mainframe computers.

Minicomputers and microprocessors provide an alternative and a complement to central processing operations. As they increase in numbers and functions, however, more time must be spent on the question of similarity and commonality of programming and operation in order to avoid a proliferation of different maintenance skills and operating systems, each requiring special attention. If distributed processing systems are to follow through on their initial favorable economics, then common protocols must be developed to minimize the cost of changes and permit the ultimate interconnection of each operating system so that the ideal of obtaining a single common information base can be attained.

Computer hardware costs are continuing to decrease. Memory costs in particular are declining rapidly, making information storage cheaper. In software, the emphasis is on simpler, less rigid programs, which permit the users to develop and operate their own systems. Ultimately, programming may disappear as a discipline, and system analysts will be system operators and users. The distinction between minicomputers and central mainframes is blurring as minicomputers get larger. Microprocessors will follow suit, closely approximating minicomputers in function and performance.

As flexible operating systems evolve, terminals will move into every department and office, providing on-line data as well as the now familiar input and output operations. Microprocessors will link together to form closely integrated systems. Problems associated with data base systems will diminish as a result of adjustment, experience, and flexibility of operating programs. The major personnel premium will be placed on people who are knowledgeable in their specialty and capable of using the computer as a tool, rather than on data processing specialists who are familiar with departmental operations.

The word processor, a combination of the typewriter and the computer, is making itself felt in office operations. The word processor takes information input on a keyboard similar to a typewriter, formats it in accordance with preprogrammed instructions, and displays the information, usually on a CRT. The information can be corrected by electronically erasing and substituting and can then be stored or printed. Printed output can serve as copy for editing, or it can be produced in a finished format. If copy is used for editing, the stored information can be recalled for further electronic editing.

The processor control provides automatic formats and the capability to add or subtract words, sentences, paragraphs, or pages in any order. This makes the word processor particularly valuable for repetitive operations that require only minor changes with each use, such as contracts, specifications, or mass mailings of notices. Revisions are handled on an exception basis, and a vast amount of material is recorded only one time.

There are other advantages as well. Information can be placed on electronic storage media, such as magnetic disks or tape, permitting automatic filing and retrieval as well as space savings. In addition, electronically stored data can be electronically transmitted, and we are just entering the era of large-scale electronic transmission of documents over long distances via telephone and radio communications paths. The present system of physically scanning pages and transmitting over telephone lines will rapidly give way to word processor–electronic communication.

The ultimate word processor will combine present functions of recording oral information on a tape recorder and transcribing taped information by manual keyboard entry by permitting direct conversion of voice input to written data. That activity, however, requires the recognition of each individual's speech patterns so that sounds can be converted to printed words, and the science of speech recognition and transcription has not yet advanced far enough to make direct transcription feasible.

Word processing represents yet another convergence of communications and data processing, further integrating the two technologies into a common entity.

IMPACTING ENERGY AND ENVIRONMENT

The revolution in electronics has the potential to moderate the constraints of our energy and environmental situation by putting us in a better position to monitor and control more carefully the interactions of each at all levels.

Automobile engine controls present a process control application that is ideally suited to electronic control. Auto manufacturers, constrained by the twin boundaries of decreased fuel consumption and decreased exhaust emissions, have turned to microprocessors to monitor engine conditions in their efforts to constrict engine operating tolerances. These microprocessors can integrate multiple inputs—engine vacuum, exhaust gas temperature and composition, and engine speed—and make adjustments to the fuel–air mixture more rapidly and carefully than a driver can. In addition, electronic measurements and controls are more reliable and require less maintenance than their mechanical, hydraulic, or pneumatic counterparts.

In a similar fashion, computers can be applied to the control of buildings. In the 1950s and 1960s, electric utilities were offering stable or decreased costs, and natural gas and fuel oil were subject to federal price ceilings. Energy was cheap and only a small part of building ownership and operating costs. Readily available energy permitted the construction of energy-intensive but simple buildings. Lots of glass could be tolerated, and insulation was not worth bothering with. But most of all, a little extra energy consumption could be traded off for simplicity in heating, ventilating, and air conditioning systems. Lighting systems that consumed a lot of energy were installed, and it was easier to let systems run 24 hours a day, seven days a week than to incorporate controls to shut them off. Air conditioning and heating were applied simultaneously as an easy solution to comfort requirements.

Rising energy costs in the 1970s caused a drastic revision in building designs. Energy became a substantial part of building operating costs.

Industry standards were modified to limit energy consumption in new buildings and equipment, and federal regulations established maximum and minimum winter and summer temperatures for public buildings. More stringent design requirements mean more careful attention to controls, and in this area, electronic processor control systems can offer substantial advantages in flexibility and reliability over pneumatic and electromechanical control systems. The building industry has been slow to respond to this new generation of control systems, but flexibility and cost are now causing a decided shift to electronic controls.

The area where electronic controls are universally accepted is building automation—the centralized supervision and control of building systems. Building automation systems provide another level of control over the local systems that control heating, air conditioning, and lighting in individual zones, allowing building operators to change system operations to meet actual building needs and providing up-to-date information about system performance and malfunctions. Automation systems are now commonly specified as integral parts of newly constructed buildings, but they can have an even greater impact on the large stock of existing buildings that have crude or limited control systems and respond poorly to attempts to decrease energy consumption. In those cases, an automation system can perform a dual function: It can provide supervisory control of local building systems that are doing an adequate job, and it can perform control functions directly in areas where controls are inadequate or nonexistent.

In concept the automation system is a system monitor that permits a building operator to trim energy use by turning equipment on or off in accordance with actual needs, rather than with simple standards, or to adjust system operating points to meter energy consumption more carefully. Examples of automation functions include controlling heating, ventilating, and air conditioning (HAC) systems and lights according to weather conditions; limiting concurrent heating and air conditioning; readjusting equipment operating temperatures to match actual heating or cooling loads due to weather and occupancy; and periodically cycling equipment off to limit excess energy inputs.

Electric utilities have been severely stressed by the cost of new power generation facilities and by stiff citizen resistance to continued rate increases. As a result, most utilities have sought or have been directed by public regulatory agencies to take maximum advantage of existing and proposed generating capacity. A utility's need for generating capacity is paced by its need to meet the peak energy requirements imposed by the customers in its service area. The almost universal use of air conditioning, which requires electrical energy, causes a substantial summer peak in utility power generation, particularly in the milder climate zones.

In order to pay the costs of meeting these peak loads and to discourage peak energy use, utilities now charge a demand charge in addition to a consumption charge. The consumption charge is based on the total amount of energy used in the billing month; the demand charge is based on the maximum rate of use during any period (usually 15 or 30 minutes) in the billing month. The demand charge is substantial and can be as much as 40 percent of the utility bill.

The demand charge creates a potential for savings that can be effectively met with a building automation system. In operation, as many of the building's electrical loads as can be turned off are connected to the automation system. The automation system monitors total energy consumption. As the demand approaches a predetermined peak, electrical loads that can be dropped are disconnected, and some mandatory loads are trimmed by the system. The net effect is to clip the peaks of energy use and fill in the valleys to provide more uniform consumption. There are simpler load management systems that only control energy use, but they lack the automation system's ability to balance electrical energy demand and occupant comfort.

Once a building energy management system is established, the control network can easily be expanded to provide access control, security, life safety, and other building management functions.

Automation systems have tracked the evolution of computer systems. In first-generation automation systems, sensors and controls were wired directly to the control unit that contained the processor. Measurements were read and control actions were taken on a time-shared basis. The second-generation equipment tied controls and sensors to local data panels, where the information was converted to a digital format and multiplexed back to the central processor. The local data panels formatted the information and made it available to the minicomputer central processor, which did all the decision making.

The latest generation of automation systems takes advantage of the ascendance of the microprocessor and distributes decision-making capability out from the central processor. This permits greater speed in decision making, builds redundancy into the system for greater reliability, and frees the central processor to do more information processing and handle larger systems with the same capacity.

One company has capitalized on technology and processing capability to develop a microprocessor/minicomputer-based system capable of monitoring and controlling all the buildings in a medium-size city through a single communications-oriented minicomputer. The system uses high-speed local data terminals connected by wire or telephone line to a powerful microprocessor that controls one or more buildings, depending on size. The microprocessor contains programs for all the equipment in its area and receives periodic updates from the central processor. Each microprocessor is tied to the central processor by dedicated telephone lines and sends digital data back to the processor as required. With much of the work load assumed by the powerful microprocessors, the central minicomputer is free to handle many buildings. Once the communications network is established within a city, multiple buildings can be electronically connected for energy management, providing a true electronic command, control, and communications network.

In the environmental area, considerable effort has been directed toward process characterization and automated control of wastewater treatment plants. Programs have been developed to characterize treatment processes and predict process control settings required to meet their effluent limitations. An evolving next step is closed loop control of process parameters. Closed loop control involves making a critical measurement, checking it against an established standard, and changing process parameters that cause the measurement to change to the desired state. Ultimately, the process computer system that controls the wastewater treatment plant can be extended to monitor the performance of the wastewater collection system, so that the entire process from the point of waste entry to the discharge of clean effluent can be monitored and controlled. This would add additional reliability to the system, making it easy to monitor and attain treatment standards.

A by-product of automated control is instant notification of problem areas in a system. Using this approach, maintenance personnel can be directed to a problem area, rather than rotated to equipment sites on a scheduled basis. This provides rapid response to specific problems and avoids the cost of maintaining continuously traveling maintenance crews. A similar approach can be taken to water distribution systems, drainage systems, and other environmental facilities within a locality.

MERGING TECHNOLOGIES

Communications and information processing are rapidly converging technologies. Digital techniques beget digital computers, which beget digital communications, which beget more digital techniques, which beget more computers. It makes sense to use the computer to provide speed and flexibility to the telephone message switching process. It also makes sense to use the same computer to convert analog signals to digital signals, and telephone companies are rapidly moving to all digital communications and switching. This digital communications capacity provides a perfect path for communications from computer to computer and from remote terminal to computer, further expanding the information processing and communications network. Voice and digital data can be transmitted simultaneously using multiplex techniques over standard conductors, coaxial cable, fiber optic cable, microwave channels, and, in the near future, laser beams.

Two-way radio systems, particularly public safety networks,

are using digital techniques to display information to field personnel over mobile digital terminals and to permit instant identification of the communicating vehicle or person in the field. Two-way channel crowding may soon be alleviated by replacing fixed channel allocation with computer-selected random channel assignment and by use of multiplexing techniques.

Satellite receiver networks and relay stations represent substantial investments in physical facilities distributed around a locality to ensure good radio reception. These facilities all tie back to a central point and, with little additional equipment investment, can be used to transmit signals from the water and sewer distribution system, provide control of traffic signals from a centralized traffic computer, monitor and control an automated wastewater treatment plant, interconnect field offices with city hall or county courthouse for voice and data communications, and provide security alarms for outlying maintenance and storage facilities, parks, recreation areas, and other facilities. With the switch from mechanical, pneumatic, and electromechanical controls to all-electronic controls, building heating and air conditioning systems can be centrally controlled, fire and life safety conditions can be monitored, and gasoline can be dispensed automatically. The possibilities are limited only by available capital and the ability to put systems to effective use.

Cable television (CATV) presents another potential low-cost communications channel. Cable TV franchises are awarded by localities to private companies for the provision of home entertainment services. The cable companies serve subscribers through a coaxial cable network with substantial signal-carrying capacity. Since the CATV companies have local franchises, communities have an opportunity during the initial granting or renewing process to include provisions for communications channels under terms that will benefit both the locality and the franchise. Most cable television operations have only one-way communications—outbound to the subscriber—but two-way technology is developing rapidly and will provide an excellent alternative path to radio channels and leased telephone lines for the transmission of voice, data, and command and control information for municipal use.

SUMMARY

The major challenge for local governments in the next few years will be to position themselves to take advantage of the time and cost savings that electronic systems have to offer. Utility crews can stop and respond on call if distribution systems and plants have information about their status telemetered to a central point. Two-way radio can dispatch the crews, and central controls can take action to minimize problems until maintenance or repairs are accomplished. License and tax sections can stop fumbling through record books and bring any citizen's status up on a screen immediately, instantly updating it when payments are made or new information is received. Planners can model the effects of new services, new growth, or zoning changes, providing better assistance to the local decision-making process.

In order to maximize the benefits, however, new skills will be required. Computers are not cheap. Electronic controls are expensive to purchase and install. Communications networks must be maintained. Cities and counties will require people with strong technical skills in electronics to assist in selecting systems that are cost-effective and complementary. Functional departments will need professionals in their disciplines who understand information and communications systems well enough to maximize their effectiveness in departmental operations. Operating systems will change, and office personnel will have to develop new skills as tasks are automated and work flows are revised. Maintenance personnel will have to add a new specialist to their complement: the communications–electronics technician.

Organizations that do develop skills to handle the revolution in electronics will be in a good position to benefit from the step increases in labor and cost savings that the revolution will produce.

CHAPTER 11

Putting It All Together

The previous chapters have focused on assessment: the need for it, tools that can be used in the process, and some discussion of key technologies that will have significant influence on our lives in the 1980s. This chapter will focus on the kinds of skills needed to conduct an assessment, how to integrate the assessment process into local government operations, and how to put the elements of assessment together in a formal technology assessment process. The chapter concludes with an example of technology assessment, and a discussion of the groundwork that was necessary to make the selected technology work.

A SEPARATE OR INTEGRAL PROCESS

Assessments are always made as an integral part of the decision-making process. Sometimes they are intuitive and cursory; sometimes they are formal and protracted. The emphasis of this chapter is therefore not so much on inventing a new complication for the decision-making process, but rather on adding formality and depth to analyzing the consequences of a decision. A more organized and in-depth analysis requires special skills—a broad knowledge of technology, an understanding of the assessment process, and some sensitivity to the decision-making process.

Depending on the needs and desires of a given community, the assessment process can be a separate function, operating apart from the other government activities, or it can be incorporated into ongoing activities. A separate assessment function is the strongest approach. The assessment staff can deal with projects on a case-by-case basis, sharpening its skills and broadening its knowledge base with each successive project. Each major project could be addressed during its formulation stages, when concepts are still fluid, affording the greatest opportunity to make corrections to eliminate negative consequences that show up in an initial assessment. The initial concept can be refined by an initial general assessment, with both the concept and the updated assessment becoming more detailed as time goes by. This permits feedback from the assessment process before concepts become so fixed that the amounts of time, money, or other resources required to effect change become so large that change is impossible.

As an alternative, the assessment process can be made an in-line part of an existing process, such as an annual budget preparation. Under these circumstances, the assessment staff will be operating under a definite time constraint dictated by the budget preparation schedule. As a result, potential targets for assessment should be identified early and prioritized to ensure that the assessment process reinforces rather than impedes the budget process. Assessment should be limited to those items that have substantial consequences and that can be evaluated before a budget decision is required. If assessments are conducted in this fashion, decisions to proceed with, drop, or defer a particular item can be based in part on the relative maturity of the technology involved and its chances for successful implementation. There is always a danger that the assessment process will be submerged in the budget process, particularly as deadlines approach and resources become scarce. For this reason, a separately identifiable assessment process is more desirable, but building the assessment process into the budgeting process is an acceptable compromise when additional personnel or a separately identifiable function cannot be justified.

WHEN TO ASSESS

A technology assessment should be conducted whenever a major capital investment is involved, particularly when there is an element of new technology; whenever a permanent facility or irrevocable change is contemplated; whenever large operational cost or large numbers of people are involved; whenever a brand new system, process, or function is being considered; and whenever rapidly changing technology is being considered. In addition, all new projects should be reviewed to determine whether such factors as ongoing support costs or rapid increase in use will make a project larger in succeeding years than it is at the outset.

STAFFING FOR ASSESSMENT

The persons conducting assessments can be full-time technology assessors, or they can be a combination of full- and part-time specialists. In any event, it is necessary to have at least one person in the organization dedicated full time to the assessment process. The assessment specialist should have a broad technical or systems background coupled with an education in engineering or the physical sciences. The required primary skill is the ability to understand the wide range of technologies that can be applied to a local government situation.

When a new function is being considered, there is a tendency to reach into the existing organization and obtain a known quantity to spearhead the operation. Local governments and public interest groups have traditionally placed people with a public administration or social science background in technology positions, feeling that the sensitivity for organizational issues is more critical than the understanding of technology. Education and experience in the social sciences, however, is generally not adequate to provide the fundamental understanding of physical principles and relationships that is necessary for a good technology assessment. Ideally, this understanding of physical principles should be accompanied by a sensitivity to what can be realistically accomplished in a local government setting.

The assessor should be a technical generalist, rather than a specialist. Assessment is not the type of job that can be performed as a first assignment out of graduate school. By the same token, it is not the type of task that should be undertaken by old warriors who have spent so much time in the trenches that they believe that nothing new can really be accomplished.

A cadre of technically trained, politically sophisticated technology specialists—variously called technology agents, change agents, or technology innovators—has developed over the last 10 years, largely because of federal programs to accelerate technology transfer. These technology transfer specialists were generally drawn from the aerospace industry, a high-technology business environment that shrank substantially in the late 1960s and early 1970s. The size and complexity of this industry developed a reservoir of people in government and industry who had an understanding of complex technology and how to develop it within an elaborate organizational structure. In addition, many of the technical advances brought on by federal investments in military and space activities were making their way into the commercial marketplace and were familiar to these technologists. These technology transfer specialists, with their understanding of complex organizations and sophisticated technology, will readily fit into the technology assessment role.

The assessment specialist should be located high enough in the organization to be free of parochial interests and must be sufficiently decoupled from day-to-day operations to be able to do the background investigation into both the technology itself and the situation in which the technology will be used. In some cases, local government size, budget constraints, or managerial structure rule out a separate function or section for technology assessment. In that case, it may make sense to team a person who is associated with, or who will use, the technology being evaluated with someone who has the skill and ability to merge the technical input into a formal assessment process. Both persons should be at the same level in the organization, have fairly broad perspectives, and be able to work together. It should be noted here that those very skills that make an excellent line manager or department head—authority, rapid decision making, pragmatism, and a concern for the here and now—are diametrically opposed to the speculative, mix-and-match, and future-oriented thinking required to conduct a good assessment. Although an assessor must keep both feet on the ground, a little willingness to consider even seemingly implausible options should be an integral part of an assessor's approach.

THE ASSESSMENT PROCESS

As previously stated, the technology assessment can be conducted on a project-by-project basis or as an integral part of another process. If a separate, project-oriented assessment is conducted, there is a nine-step procedure: initial evaluation, initial technology forecast, preliminary assessment, selection of most likely technology, detailed assessment, recommended changes, project go-ahead, monitoring and feedback, and a final retrospective assessment. The project-oriented assessment, by its nature, will emphasize the front-end assessment and choice aspects of the process. If the technology is not firmly riveted down, as it often is not at the outset, then the assessment process can unearth alternatives and present slight course corrections that can significantly benefit the project.

It would be helpful, once the project is committed, to perform Step 8 monitoring and feedback, and Step 9 retrospective assessment, in order to gain maximum benefit from the assessment process. Unless the organization is carefully disciplined, however, there is a tendency to skip in-process or retrospective activities and favor the excitement of new up-and-coming projects.

An in-line assessment generally starts after technology has been selected. Usually by the time a project has been entered for consideration in the budget process, for instance, some measurement of the technology and the scope of the activity has already been made. Therefore, the steps for an in-line assessment are initial technology forecast; detailed assessment; changes if any; a go–no-go or postpone decision; monitoring and feedback; and a final retrospective assessment. The in-line assessment forecloses some of the options and therefore some of the advantages of the project-oriented assessment, but it contains a much stronger component of in-line and retrospective assessment, since these activities can be keyed to an annual process. Particularly in the case of an assessment tied to the budget process, the assessment can be used as the basis for funding the next step.

Project Assessment

Initial evaluation. A project assessment starts with an initial evaluation of the task to be performed. The evaluation describes the nature of the task. Usually the project sponsors have a fairly good idea of the requirements and the technology necessary to support them. At this point, however, a more formal evaluation can lead to other technologies that can do the same job.

Initial technology forecast. Once one or more technologies have been defined, a simple technology forecast is made to determine the relative maturity and future path of each technology. This step sets the stage for the preliminary assessment.

Preliminary assessment. The forecast provides the basis for measuring the present and future consequences of selecting a particular technology. During this step, the favorable and unfavorable consequences are defined, using any one of the descriptive assessment techniques. If the technology is simple, one of the morphological models can be applied. If it is complex and many faceted, or if some aspects are speculative, Delphi can be used. In most cases, the analogy can be used. As a result of the preliminary assessment, the consequences can be analyzed, trading off benefits against disadvantages and mixing technologies where appropriate for added advantage.

Technology selection. At this point, the most appropriate technology is selected on the basis of its ability to fit task requirements, its relative maturity, and a preliminary assessment of the consequences of its use.

Detailed assessment. A detailed assessment of the chosen technology is made, using the descriptive tools, the analogy, or modeling techniques. The detailed assessment should present a clear picture of the consequences, both positive and negative, of adopting the technology.

Recommended changes. If the project has not proceeded too far, the requirements as well as the technology selected should be looked at again for changes that can produce greater benefit or offset some of the disadvantages. At this point, readjustments can be made with relatively little impact on the total effort. As the project proceeds and more decisions are made, more resources are committed, and changes become much more difficult.

Project go-ahead. After the assessment has been made, the consequences measured, and any adjustments introduced, an implementation decision is made. If the negative consequences are too great, even after adjustments are made, the project can be dropped, postponed, or recycled for further major adjustments in requirements or technology. If the consequences are favorable, the project can proceed to implementation.

Monitoring and feedback. During implementation, the project should be closely monitored to ensure that the results closely correlate with predictions. If they do not, then some adjustments should be made. If changes cannot be made to the technology, then the new consequences should be reassessed to ensure that they are tolerable.

Final retrospective assessment. After the implementation has been completed, actual results should be compared with predicted results. This will provide a baseline of results for future assessments.

In-Line Assessment

Initial technology forecast. This process generally starts with a defined set of requirements and a technology to meet those requirements. As a result, the chosen technology is compared with other technologies primarily to determine whether it is viable and how it will evolve.

Detailed assessment. This step is the same as for a project assessment. The most likely technology is thoroughly assessed using the techniques previously described to establish the positive and negative consequences of proceeding.

Recommended changes. The assessment will point out some areas where changes can benefit the situation. The range of changes is generally smaller than with a project assessment, since the requirements and technology have been more clearly defined.

Project go-ahead. At this point, a decision about whether to implement is made. If the negative consequences are too large, the project can be dropped or postponed. If the assessment is tied to the budget process, the results of the assessment can be used in priority setting. At this point, projects can be postponed with definite provisions for future activation in a succeeding budget year.

Monitoring and feedback. The annual budget process provides an excellent vehicle for monitoring project progress. Items in process can be targeted for an annual review, permitting small course changes or upward or downward revisions to appropriations, depending on the results obtained.

Final retrospective assessment. The annual review process also formalizes the retrospective assessment so that information can be gained for future use. This is particularly important for activities involving large investments or where a similar project is likely to be implemented in the near future.

TECHNOLOGY ASSESSMENT APPLIED TO TELEPHONE SYSTEM SELECTION

For illustrative purposes, we will trace the process actually used to select a telephone system for a county government. In this case, the formal analogy is used to assess the consequences of acquiring a telephone system from an independent supplier, an interconnect company, rather than from the local telephone operating company, which is a franchised investor-owned utility. We will evaluate each of the elements of a formal analogy— economics, management, politics, society, culture, intellectual standards, religion, ethics, and environment—in order to measure the consequences of such a decision.

Until the advent of interconnect companies, telephone systems were supplied solely by the local telephone operating company, which was an independent franchised local utility or a member of one of several regional and national communications utilities.

Since this situation is one familiar to all local governments, a previous telephone system acquisition from an operating company will serve as the base case for the analogy, and the two choices that will be compared with the base case will be the acquisition of a new system from the telephone company and the acquisition of a new system from an interconnect company.

Technology

Competing Technology

There are two major items in a telephone system. The first is the telephone itself, which converts voice to electrical signals for transmission and reconverts electrical signals back to voice signals for reception. The other major component, the switch, connects individual telephones with each other and the rest of the world. There are other components to a telephone system, but they are of lesser importance in the decision-making process.

Both the telephone operating company and the interconnect company offer switches that are comparable to previously installed systems and systems that can be considered advanced technology. The comparable technology is the crossbar system, which uses electrical relays to interconnect telephones and telephone lines in accordance with dialing instructions. These systems are mature, well proven, and reliable, and model changes are evolutionary, reflecting their advanced state of development.

Advanced technology switching systems use computers (processors) to control solid-state switches. They provide increased speed and flexibility, decreased size and power consumption, and the potential for greater reliability, since solid-state components are inherently more reliable than electromechanical devices. The use of a computer to control message paths is new, but the use of computers to make decisions and provide control functions has been thoroughly demonstrated in other applications. Thus, although the use of processor-controlled switching systems is new technology, it is a marriage of tried-and-true elements and therefore poses an acceptable risk.

The telephone company and the interconnect company can offer comparable conventional and advanced technology. For reasons delineated below under supporting technology, interconnect companies can generally offer more refined (although not necessarily better) technology.

Complementary Technology

The *Carterfone* decision in the late 1960s provided the basis for connecting telephones and switching systems of other suppliers to telephone utility trunk lines. For several years, the telephone operating companies required interface adapters to be installed between its equipment and other suppliers' equipment. Although the interface adapters protected the utility's equipment from damage from hardware whose characteristics were unknown, adapter rental fees also eroded the interconnect cost advantage, and ultimately universal standards were adopted at the direction of the Federal Communications Commission.

Under the new regulations, competing suppliers are required to register their equipment to ensure that it will be compatible with the telephone utility's equipment. This eliminates the requirement for interface adapters and guarantees that similar items of equipment will be compatible. Reputable telephone interconnect companies use only registered equipment, and the question of equipment mismatch has thus largely been eliminated. It should be pointed out that telephones themselves have long been standardized. There are ringing and wiring variations from utility to utility, but they are well recognized and accepted. Telephone variations largely consist of styling changes and convenience features, which do not affect basic system performance.

Supporting Technology

This is the area of greatest divergence between the telephone utility and the interconnect company. Interconnect companies customarily offer more variation in equipment and more convenience features. They can offer more because they have a wider range of suppliers and options and are generally small enough to be flexible. Telephone utilities, on the other hand, are generally larger and have a substantial investment in existing facilities and equipment. Because of their structure, they emphasize standardization and a relatively limited number of features and options.

The size of the telephone company's operations and maintenance force, however, assures good skills levels and the ability to respond rapidly to almost any repair problem. Interconnect companies, depending on their size and sophistication, have varying worker levels and skills but generally cannot support a labor force as large as the utility's. This is a consideration in the selection of a company. For an office operation, poor response time and limited skills may be a minor nuisance, but as the requirements move up the scale to a 24-hour-per-day emergency service operation, maintenance and repair become critical issues.

The reliability and reputation of the equipment manufacturer is another item for consideration in selecting an interconnect company. The amount of support the manufacturer is prepared to offer to the interconnect company that purchases and installs its equipment is critical. A telephone switch can be expected to have a life of 20 years or more, and a lack of spare parts or personnel if a switch is phased out, or the product line is abandoned, can severely hamper the service company and the user. One county in New York has actually experienced this situation and will eventually have to buy a new phone system since its interconnect company cannot fully maintain a switch that is now out of production. This situation usually does not occur with utilities, because their standardization and the volume of units placed in service ensure a good supply of spare parts, sometimes long after the equipment has been declared obsolete. Although some utilities provide poor service, telephone utilities generally have a service advantage, but interconnect companies can usually offer a broader range of equipment and features.

Economics

An interconnect company must be able to undersell the telephone utility. If it cannot, there is no rational motivating force to deviate from the customary service. Poor service or frustration with the bureaucracy inherent in telephone utilities may motivate a user to consider other suppliers, but in the long run, if no financial advantage is offered by the interconnect company, it is better to beg, cajole, pressure, or brutalize the utility into providing better service. There are ways to accomplish that goal, but that is not the subject of this chapter.

Before the question of economics can be addressed, we must look at the various ways that a telephone system can be provided. The most common form of service, and the one universally used before interconnect companies became a force in the market, is straight rental. Under a straight rental agreement, the user pays a monthly fee, specified in tariffs approved by the state or local regulatory body, for the use of telephone equipment. The monthly fee covers maintenance and use but customarily does not include installation or relocation.

Trunk lines, which connect the user's telephone system to the rest of the world, have traditionally been rented. This area was once the exclusive province of the telephone utilities, but recent court decisions have affirmed the right of other companies to establish themselves as telephone common carriers, and this can be expected to impact local and long-distance pricing structures.

The interconnect companies offer equipment for outright purchase or for lease. They are not governed by utility commissions and are therefore free to price equipment and service as they see fit. This lack of regulation by utility commissions also means that they are not guaranteed a profit and therefore not guaranteed to stay in business—an important consideration in the decision-making process. Interconnect companies offer outright purchase, lease with an option to buy, or straight lease for a predetermined period. Although local governments cannot derive tax benefits from lease or ownership as private companies do, their cost of capital is lower than that of business. This lowers the cost of ownership or leasing and permits localities to stabilize equipment costs, leaving only maintenance and operations costs to float with market forces.

Telephone utilities have met this competition with two alternatives to traditional equipment rental: Centrex service, where the owner pays for communications service; and the two-tier pricing system, whereby the customer pays a fixed cost that amortizes the equipment and a variable cost that pays for maintenance and operations. Under the Centrex option, owners pay a flat monthly fee for each telephone in their facilities and do not concern themselves with the details of equipment, such as trunk lines and switch gear. The telephone company is obligated to provide a level of service that meets the users' needs and makes its own decisions about the type of equipment, where it is located, and how it is serviced. This package is a service offering, not an equipment offering, and interconnect companies have no equivalent. It is available only from the telephone utility.

If the users elect to specify the type and amount of equipment they require, a competitive situation between the telephone utility and the interconnect company is created. For the usual medium- to large-size user, a PABX (private automatic branch exchange) is installed on the user's premises. For this and the other associated telephone instruments and equipment, the telephone utility offers straight rental terms or a two-tier quasi-lease arrangement. The interconnect company offers straight lease, lease purchase, or outright purchase. The two companies therefore offer either a lease or rental option or a variation of ownership. Interconnect companies have the advantage of knowing the utility's tariffs, and they have the ability to price lower.

Given the total life span of a telephone system, initial cost is only part of the total cost of operation. Capital cost, maintenance cost, operations cost (including trunk line rental and telephone operators' salaries), and the cost of moves and service changes have to be estimated over the useful life of the equipment in order to ensure that total life cycle costs are compared. If some caps can be placed on maintenance and service charges, the interconnect company should have a cost advantage over the telephone utility.

Management

A telephone utility contract is easy to manage. Rates are fixed by tariff, and ordering and changing equipment can be done on an informal basis if necessary. On the other hand, telephone utilities tend to be bureaucratic, and it may take substantial effort to get a properly motivated sales installation and maintenance force to provide responsive, efficient service. By contrast, the working arrangement with an interconnect company is a contractual situation and requires a good contract manager to devote at least part of his time and energy to telephone services. Unless service rates and additional equipment costs have been specified in the original agreement, each telephone move or equipment change is a separate contract negotiation. Most organizations have frequent departmental shifts and office changes driven by growth, shrinkage, or changing priorities, and modifying telephone service can get expensive if changes are frequent. The last few years have been unstable in terms of inflation, wage rates, and taxes, and few private organizations can be expected to enter into long-term fixed-rate service agreements. Periodic maintenance contract renewal negotiations will be necessary.

Processor-controlled systems provide the capability to change some telephone numbers and features by reprogramming rather than by rewiring. They also provide integral fault isolation capability. This enables users to perform some minor maintenance and relocation tasks with their own staffs at reduced cost. This benefit must be weighed against the effort required to hire, retain, and manage a skilled staff capable of performing those functions.

Overall, a responsive telephone utility is easier to manage than an interconnect company.

Politics

The telephone utility is a strong and visible operation within a community. The utilities are not afraid to use their influence with elected officials if they feel that contracting for service with an interconnect company will hurt their status or impact their revenue base. Since a large change in service and a change from the traditional way of doing business must be reviewed by the governing body, some careful measurements of the elected officials' attitudes toward contracting with an interconnect company are required. To the extent possible, the utility's attitude should also be measured. When the case for a decision is presented, it should be based on absolutes, such as decreased cost or better communication, so that the decision is based on business imperatives and cannot be construed as an unjustified attack on the utility.

One other issue should also be considered. Interconnect companies are free to pick and choose their customers; utilities are required by law to serve all customers. Some thought should be given to the impact that a change to an interconnect company will have on the utility. Although the telephone utility is chartered to make a profit, it is a public-benefit corporation with a role much akin to a local government's in its specialized area. To the extent that the telephone utility suffers, local citizens suffer, and it is in a locality's interest to do what it can to ensure a healthy telephone utility.

Society and Culture

Telephones are a well-accepted part of our society, and telephone service is a necessary medium of communication between citizens and government. Consequently, a change in telephone service does not have substantial societal and cultural overtones. There may be some positive response to the concept of the county reaching for greater efficiency and more effectiveness by changing from the traditional utility service to a newer free market supplier, and this could result in a more positive image for the county in the eyes of its citizens. However, this was not an important ingredient in the assessment.

Intellectual Standards

A fundamental change in a communications system will affect all its users. Therefore it is well to involve opinion leaders and decision makers, such as department heads and key officials, early in the process, pointing out the benefits to them and the public so that they are committed to the change. If their support is obtained, it will heavily influence the people who will use the system, and will help to overcome the cumulative individual resistance that can easily sink a project.

Religion and Ethics

If the decision to purchase a system from a private supplier is couched in terms of cost and efficiency, then there will be little, if any, public resistance. However, if the decision is based on vague presentments of poor service, being "ripped off" by the same rates the rest of the community is paying, or other similar complaints, then the decision will be viewed as an attempt to treat the telephone utility unethically, and questions of integrity will be raised. Objective reasons for change will substantially blunt this potential criticism. The option of choosing an interconnect company is rooted in the Federal Communications Act, and businesses and communities are free to choose the option that provides them with the best service. There was no perceived problem in the county in this area.

Environment

Telephone service involves electronic and electromechanical equipment, which has negligible effect on the environment.

Results

As a result of the assessment, the county elected to acquire an advanced electronic switching system from the telephone utility under the two-tier quasi-lease arrangement. The combination of a lump sum initial payment and the features of an advanced switching system allowed for greatly improved service at a very small increase in cost and put a cap on part of the system cost. The telephone utility was chosen over the interconnect company because none of the local interconnect companies had installed a system of comparable size in the service area. In addition, the switching system served five buildings in a government center, including a public safety building with a full-time public safety communications center, and it needed the kind of instant maintenance response that could only be guaranteed by the telephone utility. One interconnect company offered a switching system that was more advanced and had been in service longer than the one provided by the utility, but the advanced technology was more than the county needed, and for that reason, it was not selected.

In subsequent competitive procurements of telephone systems for a nursing home and an office building, the county selected a reputable interconnect company over the telephone utility. The interconnect system provided a cost-effective alternative for these smaller, less demanding system requirements.

THE OTHER SIDE OF ASSESSMENT

The preceding example outlined the analytical process used to arrive at a decision in the selection of a telephone system. There was also a softer side—ensuring that once a technology selection was made it would work.

The first step in that process was an actual visit to the switching system supplier's engineering and manufacturing facilities to ensure that the representations of their sales and marketing force were true and that the systems were mature enough to guarantee trouble-free installation and operation.

The new telephone system also represented a radical departure from the old system. One centralized system would be installed in place of many individual departmental systems, and the apparent flexibility of multibutton telephones would be replaced by plain telephones tied to a sophisticated switching system. This approach was cost-effective, but its virtues had to be sold. In order to ensure acceptance, presentations were made to each operating department head in the county government structure, explaining the limitations of the existing system and how they would be overcome by the new system. The apparent advantages of the old system were placed in a true context, and the major features of the new system—better communications at lower cost—were outlined. The managers concurred with the concept, and their concurrence was followed up with a detailed review of each department's needs. The review accomplished two things: It tailored the telephone system to each department's needs, and it prepared personnel for the change.

It was also necessary to convince the local telephone operating company that early planning was required to ensure a smooth transition and good customer acceptance. The operating company recognized its competitive situation with an interconnect company and was willing to devote additional resources to this situation to enhance its position.

Potential interconnect companies were narrowed down to one that had the resources to compete with the local telephone operating company. The competitive situation, including prospects for sales to the county, was outlined clearly at the outset to the interconnect company. The interconnect company provided valuable input to the design and procurement process, and although it did not win this competition, it felt that it had been treated fairly. This company subsequently sold systems to the county for other buildings.

There were compromises made in the system design. The building inspection department was under considerable external and internal pressure during that period, and the departmental telephone system was deliberately overdesigned in order to improve public access to inspectors and other personnel. During the actual telephone installation, the county judges reversed their previous position and insisted on a multibutton telephone system. In view of their place in the government structure, there was no choice but to provide what they felt they needed.

Before the systems were installed, each department received training in their use from an experienced telephone company representative to ensure a minimum of interruptions in communications during the transition to new offices.

From the outset, it was recognized that some telephone system features were too cumbersome and too sophisticated to be widely used. They were an integral part of the system package, however, and were available to those people who wanted to take advantage of them. Thus, although some of the technology was more than was needed, it did not conflict with the basic system used and was therefore acceptable.

The true measure of successful technology assessment is how the technology works in its setting. Although selecting technology that is mature and trouble free is essential to the process, it is not enough. The technology must be appropriate, it must be acceptable, and it must complement the social environment in which it functions.

Index

Aberdeen Proving Ground, and
computer development, 164
aerospace industry, and technology forecasting, 18
agricultural fertilizers, side
effects of, 13
air pollution, as result of automation, 15
AM (amplitude modulation) radio
frequency, 162
American Water Works Associations (AWWA), on GAC fil-
ter, 134
Amtrak, 151
analog computer, development of,
164
analogy
definition of, 81
formal, 82–96
as technique, 80–81
anonymity, in Delphi process, 53
approximations, as model limitations, 97
Arab oil embargo, effect of, on
energy situation, 140
Armstrong, E. H., 162

assessment(s)
in decision-making process,
185–186
in-line, 189, 191–192
process of, 189–192
project-oriented, 189, 190–
191
staffing for, 187–189
timing of, 187
assessor, qualities of, 188
AT&T, and communications industry, 168
auto emission inspection programs, 125–126
auto emission tests, implementation barriers of, 136
automation systems, use of, 179–
181
Automative Age, 141
automobile, effect of, on society,
14–16
automobile industry
effect of Clean Air Act on, 127–
128
as reflecting energy conservation, 157

Babbage, Charles, and computer development, 163
Battelle Columbus Laboratories
and interpretive structural modeling, 100
and National Science Foundation experiment, 88
Bell, Alexander Graham, 161
Bell Laboratories, and transistor development, 162
"best available technology" (BAT), and private treatment plants, 121
biomass, as energy alternative, 154–156
breakthroughs, effect of, on technology forecasting, 19
breeder reactors, as energy source, 152–153
building automation, and computers, 179–180

cable television (CATV), development of, 183
carbon dioxide (CO_2), and fossil fuel combustion, 147–148
Carterfone decision, 170, 194
cathode-ray tube (CRT)
in communication system, 173
in data processing, 175
Census Bureau, and data bases, 112
Centrex service, and telephone systems, 196–197
chemicals, effect of, on society, 13–14
Chrysler Corporation, and auto industry losses, 128
cities
effect of automobile on, 15
as local government, 27
see also local government
Clean Air Act, 116, 124–129
Clean Water Act, 133

"closed loop control" process, as automation, 181–182
coal, as energy alternative, 150–152
combined sewer overflows, as environmental problem, 122
combustion products, as pollution, 15–16
communication(s)
computers and, 163–166
data processing and, 174–178
as industry, 161–162
local government needs and, 166–168
radio and, 162–163
comparison, in formal analogy, 83
Comprehensive Employment and Training Act (CETA), 39
computer(s)
and automobile industry controls, 178
and building controls, 178
and communication, 163–166
in decision-assisting model, 101
in modeling, 98
Conrail, 151
conservation, of energy, 156–158
corn, as biomass energy source, 155
cost reduction, as economic recovery, 86
costs, of growth, 110
Council on Environmental Quality, on hazardous wastes, 132–133
counties
as local government, 26
see also local government
crossbar switch, in communication system, 169
cultural values
as subject in assessment process, 199
and technology, 92–93

data bases
DIME (Dual Independent Map Encoding) file, as, 112
data points, and measurement parameters, 50
data processing
in communication system, 174–178
digital computer in, 174
distributed, 175
microprocessor in, 175–176
minicomputer in, 175
and payroll system, 174
word processor in, 176–177
Davis-Bacon Act, 32
DDT, and side effects of technology, 72
decision-assisting models, 100–102
decision-making process, forecasting in, 20–21
deForest, Lee, 161
Delphi process, 53–59
demand charges, use of, 144, 180
demographic models, 100
detailed assessment
in in-line assessment, 191
in project-oriented assessment, 190
devaluation of dollar, effect of, 146
diesel engine, effect of Clean Air Act on, 128
differential analyzer, development of, 164
digital computer
in data processing, 174
development of, 164
Digital Equipment Corporation, and computer development, 165
DIME (Dual Independent Map Encoding) file, 112

diode, as electronic tube, 161
direct expenditures, effect of, on economy, 37
distributed data processing, 175
dollar, devaluation of, effect of, 146

Earth Day, 116
Economic Development Administration (EDA), 39
economics
as subject in assessment process, 195–197
and technology, 86–87
Edison, Thomas, 161
education, as local government service, 27–28
EDVAC (Electronic Discrete Variable Automatic Computer), 164
election cycle, two-year, as barrier to innovation, 34–35
election results, and predictive techniques, 99
electronics, and energy, 178–182
emission (auto) inspection program, 125–126
employment, as measure of federal government growth, 38
energy
effect of automobile on, 16
and electronics development, 178–182
energy alternatives
biomass energy as, 154–156
coal as, 150–152
natural gas as, 148–150
nuclear power as, 152–153
solar power as, 153–154
synfuels as, 156
ENIAC (Electronic Numerator, Integrator, and Computer), 164

environment
   as element in formal analogy,
      95–96
   as subject in assessment pro-
      cess, 200
Environmental Defense Fund
   (EDF), on GAC filters, 134
environmental impact statement
   (EIS), 118
Environmental Protection Act
   (EPA), 33
   and clean water, 133
   and resource recovery, 131–132
ethics
   as element in formal analogy,
      94–95
   as subject in assessment pro-
      cess, 199–200
evaluation
   in formal analogy, 83
   in project-oriented assessment,
      190
   as step in decision-making pro-
      cess, 21
expert forecaster, use of, 52–53
"expert opinion" approach, as in-
      formal forecasting method,
      44
extrapolation
   limits of, 51–52
   as predictive tool, 44–45

facility-oriented services, and
      financial models, 110–111
Federal Communications Com-
      mission (FCC), 168
Federal Environmental Pesticide
      Control Act, 117
federal expenditures, 37–38
federal funding, requirements for,
      132–133
Federal Insecticide, Fungicide
      and Rodenticide Act (FIFRA),
      117

federalism, effect of, on innova-
      tion, 35
federal powers, flow of, 32
feedback, in Delphi process, 53
financial forecasting models, 100
financial models, 109–111
fish bowl environment, as barrier
      to innovation, 36–37
"fish-swimmable," as water pollu-
      tion goal, 120
Fleming, Sir John, 161
floppy disk, development of, 166
flow diagrams, as tool of norma-
      tive forecasting method, 74–
      79
flow of power, in government, 32
FM (frequency modulation) radio
      frequency, 162
food production, effect of technol-
      ogy on, 12–13
Ford Motor Company, and auto
      industry losses, 128–129
forecast
   in in-line assessment, 191
   in project-oriented assessment,
      190
forecasting methods
   and breakthroughs, 19
   and decision-making process,
      20–21
   hazards of, 19–20
   informal, 42–44
   normative methods of, 60–79
   predictive tools for, 44–57
   in social system, 21–24
   technology, 16–19
formal analogy
   as assessment tool, 82–83
   economics, as element in, 86–
      87
   environment, as element in,
      95–96
   intellectual standards, as ele-
      ment in, 93–94

management, as element in, 88–90

politics, as element in, 90–92

religion and ethics, as elements in, 94–95

society and culture, as elements in, 92–93

steps in, 83

technology, as element in, 86

fossil plants, as energy source, 144

fusion, nuclear, as energy source, 153

GAC filter system, 138–139

gas, natural, as energy alternative, 148–150

gasification, as energy alternative, 152

gasoline engines, effect of Clean Air Act on, 127–128

General Electric Company, 161–162

General Motors Corporation, and auto industry changes, 129

GNP (gross national product), and government expenditures, 37

government
  expenditures of, 37
  and forecasting, 22–24
  see also local government

granulated activated carbon (GAC) filters, 133–134

growth, costs of, 110

hazardous waste, treatment of, 132–133, 138

Hertz, Heinrich, 161

Hollerith, H., and punch card tabulation, 164

hospitals, and energy conservation, 157

human element, in formal analogy, 83–84

human services, by local government, 27

hydrological model, and physical simulation, 103

hydropower, as energy source, 144

IBM, and computer development, 165

implementation
  as decision-making process step, 21
  in in-line assessment, 192
  in project-oriented assessment, 191

incineration, as resource recovery alternative, 137

industrialization, effects of, 12–13

industry, and forecasting, 22

information processing, by local governments, 167

in-line assessment, process of, 191–192

innovation, barriers to, 30–37

Intel Corporation, and computer development, 166

intellectual standards
  as element in formal analogy, 94–95
  as subject in assessment process, 199

intergovernmental transfer, as federal expenditure, 37–38

interpretive structural modeling, 100–101

James River, and side effects of technology, 12–13

kepone, and side effects of technology, 12–13

landfilling, as resource recovery alternative, 137
land use
 as barrier to innovation, 30–31
 controls, 121
legislation, of local governments, 27
Lincoln Laboratories, and computer development, 165
liquification of coal, as energy alternative, 152
local government
 and automation, 167
 cities as, 26
 communication needs of, 166–168
 counties as, 26
 definition of, 25
 and education, 27–28
 energy use of, 143–144
 power of, 32
 role of, 25–27
 as service delivery corporation, 27–29
 telephone use by, 167
 two-way radio use by, 167
long-range decision making, and two-year election cycle, 34–35
long-range planning, distrust of, 39

management
 as subject in assessment process, 197–198
 and technology, 88–90
Marconi, 161
Mark I, and computer development, 164
Marine Protection Research and Sanctuaries Act (Ocean Dumping Act), 117
Martino, Joseph (*Technology Forecasting for Decision Making*), 98

mass transit system, effect of automobile on, 15
maturity levels, and measurement parameters, 50–51
measurement parameters, 49–51
mechanization, effect of, on food production, 12
microcircuit, development of, 162
microcomputer
 in data processing, 175–176
 development of, 166
microprocessor, development of, 166
military decision making, and technology forecasting, 17–18
minicomputer
 in data processing, 175
 development of, 165
mining, coal, 151
mobile emission requirements, 127
mobility, as result of automobile, 14
models
 computers and, 98
 data bases for, 111–115
 decision-assisting, 100–102
 definition of, 97
 demographic, 100
 financial, 109–111
 financial forecasting, 100
 hydrological, 103
 physical simulation, 100, 102–108
 population projection, 100
 quality, 104
 quantity, 103–104
 service delivery, 100, 108–109
 value of, 97
moderator, in Delphi process, 54–55
monitoring
 in in-line assessment, 192
 in project-oriented assessment, 191

Moore School of Electrical Engineering, and computer development, 164
morphological models, as tool of normative forecasting method, 67–73
Morse, Samuel, 161

National Advisory Commission on Aeronautics (NACA), and technology forecasting, 18
National Aeronautics and Space Administration (NASA), and forecasting, 43
National Criminal Information Center (NCIC), 173
National Environmental Policy Act, 116, 118
National Science Foundation, 88
natural gas, as energy alternative, 148–150
Natural Gas Policy Act of 1978, 149
node, as element of relevance tree, 61
Noise Control Act, 117
normative methods, of forecasting definition of, 60
  flow diagrams, as tool for, 74–79
  morphological model, as tool of, 67–73
  relevance trees, as tool of, 61–67
nonmonetary benefits, 87
nonmonetary costs, 86
nonpoint sources, and pollution control, 120
North Slope oil, 148
"nothing is changing" approach, as informal forecasting method, 43–44
NOX emissions, 128
nuclear plant development, 147

nuclear power, as energy alternative, 152–153
Nuclear Regulatory Commission, 153

oil
  Arab embargo and, 140
  consumption, 16
  environmental problems and, 147
  history of, 141–143
  imports, 142
OPEC, 140, 145
opinion makers, role of, in technology acceptance, 93–94
organizational structure, as barrier to innovation, 35–36
orientation, in Delphi process, 54
outer continental shelf (OCS) exploration, 148–149
Oxford English Dictionary
  on analogy, 81
  on local governments, 25
  on models, 97

PABX (private automatic branch exchange), and telephone system, 197
parameters, measurement of, 49–51
passivity, of local government, as barrier to innovation, 31–34
Pearl, Raymond, 46–47
Pearl curve, as predictive tool, 45–48
phonograph, and communication, 161
physical science elements, in formal analogy, 83
physical simulation models, 100, 102–108
planning, as step in decision-making process, 21

point-for-point correlation, and formal analogy, 82
point sources of pollution, controlling, 120
politics
  as subject in assessment process, 198–199
  and technology, 90–92
population projection models, 100
prediction, in formal analogy, 83
predictive tools
  extrapolation as, 44–45
  Pearl curve as, 45–48
  trend curve as, 48–49
preliminary assessment, in project-oriented assessment, 190
President's Urban Policy Program, 131
price ceilings, and domestic oil production, 149
primary pollution standards, 124
processor-controlled switch, in communication system, 169
project-oriented assessment, process of, 189, 190–191
Public Law 92-500, 116–117, 119
public safety, and local governments, 27
public school system, and local governments, 27–28
Public Technology Incorporated, in National Science Foundation experiment, 88

quality standards, and Clean Air Act, 126–127
questionnaire, in Delphi process, 53, 56–57

radio dispatch center, computers and, 172
radio frequencies, 162–163
railroads, and coal as energy source, 151

Rand Corporation, and Delphi process, 53
Random Access Memory, development of, 166
readjustments
  in in-line assessment, 191–192
  in project-oriented assessment, 191
relevance trees
  node, as element of, 61
  as normative forecasting method tool, 61–67
  weighted, 65–67
religion
  as element of formal analogy, 94–95
  as subject in assessment process, 199–200
Resource Conservation and Recovery Act (RCRA), 117, 129–133
resource recovery facilities, 131
resource recovery programs, use of, 136–137
retrospective assessment
  in in-line assessment, 192
  in project-oriented assessment, 191
return on investment, providing, as economic criteria, 86–87

Safe Drinking Water Act, 117, 133–134
satellite receivers, in communications system, 172
schools, and energy conservation, 157–158
secondary pollution standards, 124
semiconductor, development of, 162
service delivery corporation
  local government as, 27–29
  technology and, 29–30

service delivery models, 100, 108–109

Siemens, E. W., 161

Sierra Club, and Clean Air Act, 125

skills, support, as management consideration, 88–90

social-physical systems, technology prediction in, 98

society
  and automobile, 14–16
  as subject in assessment process, 199
  and technology, 12–14
  values of, and technology, 92–93

Sohio Company, and oil pipeline, 147

solar power, as energy alternative, 154

Sperry Rand Corporation, formation of, 165

staff
  for assessment, 187–189
  of local government, 28–29

state implementation plans (SIPs), 124

state-of-the-art measurement, and measurement parameters, 50

stationary source emission requirements, 127

stimuli, design, in physical simulation model, 102–103

stock market, as reflecting society as evolving, 21–22

strip mining, problems of, 151

structure, organizational, as barrier to innovation, 35–36

suburbs, as result of automobile, 15

sugar, as biomass energy source, 156

sunshine laws, as barrier to innovation, 36

switching systems, in communication systems, 168–169

synfuels, as alternative energy source, 156

taxes, as measure of federal government growth, 38

technology
  in analogy construction, 84–86
  assessment, 17
  as assessment factor, 82
  competing, 84
  complementary, 84, 85
  culture and, 92–93
  definition of, 17
  economic considerations of, 86–87
  environment and, 95–96
  forecasting, 17–20
  impact of, 88–90
  intellectual standards and, 93–94
  and management, 88–90
  as merging, 182–183
  politics and, 90–92
  primary, 85
  religion and ethics and, 94–95
  and service delivery, 29–30
  side effects of, 12–13
  and social-physical system, 98
  society and, 12–14, 92–93
  as subject in assessment process, 193–195
  supporting, 84, 85
  support skills for, 88–90

Technology Forecasting for Decision Making (Martino), 98

technology selection, in project-oriented assessment, 190

telephone system
  as communication, 161, 168–171
  computers in, 169–170
  use by local government, 167

Three Mile Island, 153
time division multiplexing, in communication system, 169
time span abbreviation, as model use, 97
tipping fees, in solid waste recovery program, 131
Toxic Substances Control Act, 117
traffic control measures, implementation of, 136
transfer payments, effect of, on economy, 37
transistor, development of, 162
trend curve, as predictive tool, 48–49
two-way radios
    as communication systems, 171–172
    and local government, 167
    and technology, 182–183
two-year election cycle, as barrier to innovation, 34–35

UHF (ultrahigh frequency) radio frequency, 163
Union Electric Company, and resource recovery facilities, 131
UNIVAC (Universal Automatic Computer), 164–165
U.S. Department of Energy, and resource recovery, 132
U.S. Environmental Protection Agency, 117
vacuum tube, and transistors, 162
VHF (very high frequency) radio frequency, 163
video communication, 162

waste
    hazardous, treatment of, 132–133
    solid, treatment of, 130–132
waste dumps, cleanup of, 129
Water Pollution Control Act, 117, 118–124
water quality standards, establishment of, 120
Webster's Seventh New Collegiate Dictionary, on technology, 17
weighted relevance trees, 65–67
    see also relevance trees
"window blind" approach, as informal forecasting method, 42–43
wood, as biomass energy source, 155
word processor, in data processing, 176–177

"zero discharge," as water pollution goal, 120
zoning
    as barrier to innovation, 30–31
    as local government service, 23